PowerPC Programming
Pocket Book

About the author:

Through his work with Motorola Semiconductors, the author has been involved in the design and development of microprocessor based systems since 1982. These designs have included VMEbus systems, microcontrollers, IBM PCs, Apple Macintoshes, and both CISC and RISC based multi-processor systems, while using operating systems as varied as MS-DOS, UNIX, Macintosh OS and real time kernels.

PowerPC Programming Pocket Book

Steve Heath

Newnes
An imprint of Butterworth-Heinemann
Linacre House, Jordan Hill, Oxford OX2 8DP
A division of Reed Educational and Professional Publishing Ltd

ℛ A member of the Reed Elsevier plc group

OXFORD   BOSTON   JOHANNESBURG
MELBOURNE   NEW DELHI   SINGAPORE

First published 1995
Reprinted 1996

British Library Cataloguing in Publication Data
A catalogue record for this book is available
from the British Library

ISBN 0 7506 2111 7

Typeset by *Steve Heath*
Printed and bound in Great Britain by Clays Ltd, St Ives plc

Contents

Appendices

Preface

The PowerPC architecture is the most successful RISC architecture that has yet appeared with many adopters ranging from Apple and IBM in the personal computer market to Ford for its power train control in its car and truck products. The RISC architecture has moved out of the laboratory and niche markets and applications into the mainstream computer arena.

As a result, many within the software industry are faced with the task of learning new programming techniques and more importantly which instructions are generic and which are specific to a particular processor. With the documentation associated with the PowerPC architecture and the processors occupying several feet of shelf space, this work provides the essential information about the programming models for the MPC601, MPC603 and MPC604 PowerPC processors and the instruction sets for all the processors including the 64 bit varients and alternative mnemomics in a compact and less formal format.

The first chapter describes the generic programming model which is common to all the processors within the PowerPC family. The subsequent chapters describe the programming models specific to the MPC601, MPC603 and MPC604 32 bit PowerPC procesors. The next chapter describes the exception model and how it works and the final chapter describes the instruction set, and addressing modes, including the alternative mnemonics and 64 bit instructions. The five appendices describe which instructions are supported by which part of the architecture, a M68000 to PowerPC instruction cross reference and special purpose register numbering.

I would like to express my thanks to Motorola for their encouragement and help and once again, to Sue Carter for her support and editing skills.

Steve Heath

Acknowledgments

The material in this book has been based upon PowerPC: a practical companion (Butterworth-Heinemann ISBN 0-7506-1801-9).

By the nature of this book, many hardware and software products are identified by their tradenames. In these cases, these designations are claimed as legally protected trademarks by the companies that make these products. It is not the author's or the publisher's intention to use these names generically, and the reader is cautioned to investigate a trademark before using it as a generic term, rather than a reference to a specific product to which it is attached.

All trademarks are acknowledged, in particular:

- IBM, IBM PC, PC XT, PowerPC, PC AT and PC-DOS are trademarks of International Business Machines.

- Motorola, MC68000, MC68020, MC68030, MPC601, MPC603, MPC604, MPC620, MC68040, are all trademarks of Motorola , Inc.

While the information in this book has been carefully checked for accuracy, neither author nor publisher assume any responsibility or liability for its use or any results, or any infringement of patents or other rights of third parties that would result.

As technical characteristics are subject to rapid change, the information contained is presented for guidance and education only. For exact detail and design, always consult the manufacturers' data and specifications.

Many of the techniques within this book can destroy data and such techniques must be used with extreme caution. Again, neither author nor publisher assume any responsibility or liability for their use or any results.

Chapter 1
The PowerPC programming model

This chapter describes the PowerPC programming model and how it is implemented. Like most CISC processors, the model is divided into two: a user level or mode where typically tasks and programs execute and a supervisor level or mode which has access to all the user level registers as well its own, which are used to handle interrupts and housekeeping, and so on. Typically, operating system software executes in the supervisor mode.

There are other distinctions: generally, the user mode is consistent across the family (the MPC601 is an exception to this) but the supervisor mode is processor dependent. The only way to move from the user level to the supervisor level is if an exception (an interrupt, error, trap or system call instruction) occurs. Normally, the rfi instruction is executed to get back from the supervisor mode.

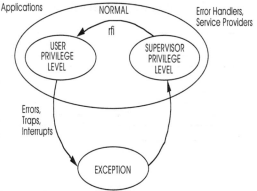

The exception transition model

The PowerPC user mode register set

The PowerPC user mode register set consists of 32 x 64 bit general purpose registers which can be used for holding data, as part of address calculations, stack pointers and other functions associated with data and address registers. These registers are used to store integer values and are used in integer and logical operations.

To support floating point calculations, there are 32 x 64 bit floating point registers that support double precision arithmetic.

The link and count registers are 64 bit registers used to store target and return addresses for some branch instructions. There is also a 32 bit condition register which is split into eight smaller 4 bit condition registers so that multiple conditions can be stored.

General Purpose Registers

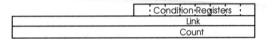

Branch Control Registers

Floating Point Registers

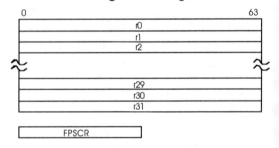

Logical Memory Address Space

PowerPC programming model

The last two registers are the floating point status and control register, FPSCR, and the integer exception register, XER, which contains carry and overflow flags and is used to store count values and other information for the string search instructions.

The diagram shows the PowerPC architectural definition and although it defines most of the registers as being 64 bit and capable of storing a 64 bit double word, the implementation in the MPC601, MPC603 and MPC604 is the 32 bit derivative. With the 32 bit version, the general purpose registers, the link and condition code registers, FPSCR and XER are all truncated to the lower 32 bits to form a 32 bit wide programming model. This allows the programming model to be extended or fully implemented in a 64 bit PowerPC processor while still retaining compatibility with the 32 bit implementation. All the control and status bits that are in the XER and FPSCR are in the lower 32 bits and are therefore consistent with both 32 and 64 bit versions.

As the majority of available PowerPC processors and systems are 32 bit implementations, the 32 bit model will be used throughout this book, unless specified otherwise.

Please note: unlike most CISC processors, there is no program counter within the PowerPC architecture. This is not an omission. The reason is that it becomes difficult to control with superscalar processors that execute multiple instructions per clock cycle. The main issue concerns which instruction is actually the program counter — especially with out-of-order execution.

General purpose registers

The 32 general purpose registers are interchangeable, with the exception of r0. This register can take the value of zero in certain cases and this is used to synthesise data moves, for example. To move data from register r4 to r6, the add immediate instruction is used where zero — register r0 — is added to r4 and the result is stored in r6. The effect is to copy the contents of r4 and place it into r6. The rules for r0 are as follows:

- If r0 is used as part of an effective address calculation, its value is taken as zero.
- If r0 is used with either the add immediate or add immediate shifted instructions, its value is taken as zero.
- In all other cases, r0 operates like a normal general purpose register.

With the 32 bit implementation, the load and store double word instructions are considered illegal and force an exception. The supervisor could simulate the access, if needed, but it would have to use two general purpose registers to hold the value which is a different result from using a single 64 bit register.

Floating point registers

The 32 floating point registers are 64 bits wide in both the 32 bit and 64 bit programming models and there are special floating point load and store instructions to support them.

S	EXP	FRACTION
0 1	8 9	31

Floating point single precision format

S	EXP	FRACTION

0 1   11 12   63

Floating point double precision format

Condition register

The condition register is 32 bits wide but it is divided into eight smaller condition registers of four bits numbered CR0 to CR7. When they are used to store the results of a compare instruction, each one has less than, greater than and equals flags, and the flag for overflow or unordering, depending on whether the comparison is on integer or floating point values.

CR0 and CR1 can be updated automatically by using the record option for most instructions. This is done by adding a full stop or period after the instruction mnemonic. If the instruction involves an integer operation, CR0 is updated with less than, greater than, equals and overflow status. If the operation involves floating point, CR1 is updated by copying bits 0 to 3 from the FPSCR.

Condition register CR

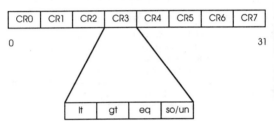

CR0	CR1	CR2	CR3	CR4	CR5	CR6	CR7

0   31

lt	gt	eq	so/un

The condition register

Floating point status and control register

This is a complicated register that provides status information and allows the user to customise floating point behaviour.

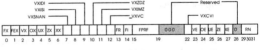

Floating point status and control register — FPSCR

Flag	Description
FX	Floating point exception flag. Set by every floating point instruction that causes an exception. This is a sticky bit and must be cleared by software.
FEX	Floating point enabled exception flag. Set by every floating point instruction that causes an enabled exception condition. If this is clear and FX is set, then the exception is one that the user has no control over. If both FX and FEX are set, then the

	exception has been caused by an exception that the user has explicitly enabled.
VX	Floating point invalid operation exception flag. This is set if the exception cause was an invalid operation.
OX	Floating point overflow exception.
UX	Floating point underflow exception.
ZX	Floating point zero divide exception.
XX	Floating point inexact exception.
VXSNAN	Invalid operation involving a SNaN. This is a sticky bit and must be cleared by software.
VXISI	Invalid operation involving infinity minus infinity. This is a sticky bit and must be cleared by software.
VXIDI	Invalid operation involving infinity divided by infinity. This is a sticky bit and must be cleared by software.
VXZDZ	Invalid operation involving zero divided by zero. This is a sticky bit and must be cleared by software.
VXIMZ	Invalid operation involving infinity divided by zero. This is a sticky bit and must be cleared by software.
VXVC	Invalid operation involving an invalid compare. This is a sticky bit and must be cleared by software.
FR	The floating point fraction has been rounded.
FI	The floating point fraction is inexact.
FPRF	Floating point result flags.
VXSOFT	Floating point invalid operation. This flag allows software to set a invalid operation condition that may not be associated with floating point operations such as the square root of a negative number. Not implemented in the MPC601.
VXSQRT	Invalid operation involving square root. This is a sticky bit and must be cleared by software. This is not implemented in the MPC601.
VXCVI	Invalid operation involving integer conversion. This is a sticky bit and must be cleared by software.
VE	Enable invalid operation exceptions.
OE	Enable overflow exceptions.
UE	Enable underflow exceptions.
ZE	Enable zero divide exceptions.
XE	Enable inexact exceptions.
RN	Floating point rounding control
	00 round to nearest
	01 round towards zero
	10 round towards +infinity
	11 round towards -infinity

Floating point status and control register flags

Result flags (bits 15-19)	Description
10001	Quiet NaN
01001	- infinity
01000	- normalised number
11000	- denormalised number
10010	- zero
00010	+ zero
10100	+ denormalised number
00100	+ normalised number
00101	+ infinity

FPSCR FPRF result flag settings

Integer exception register

This register is used for two functions: it provides the carry and overflow flags together with a summary overflow flag and stores the compare byte and byte count number used by some of the string handling instructions.

SO	OV	CA	0000000000000	compare byte for load string & compare byte instruction	0	number of bytes for load/store string instrs.
0	1	2	3 15	16 23	24	25 31

▨ Reserved

SO	Summary Overflow	Sets when OV sets. Is a sticky bit (software must clear it).
OV	Overflow	If Overflow is enabled in the opcode (OE=1), OV sets if the instruction overflows (assumes signed numbers, i.e. the carry out of bit 1 is not equal to the carry out of bit 0).
CA	Carry	Sets if there is a carry out of bit 0 if the instruction is a "Carrying" instruction (ends in "c") or an "Extended Precision" instruction (ends in "e"). Extended Precision instructions also use CA as an operand.
bits 16-23		Contains byte to be compared in the lscbx (load string & compare byte) instruction.
bits 25-31		Contains the byte count for lscbx, lswx, and stswx instructions.

Integer exception register — XER

This is a special purpose register accessed from the user level by the 'move to and move from special purpose' instructions.

Link register

The link register is used to hold 32 bit target addresses for use with certain branch instructions. The two least significant bits of this register are effectively ignored when its contents are used as an address.

This is a special purpose register accessed from the user level by the 'move to and move from special purpose' instructions.

Count register

The count register is used with certain branch instructions either to hold a count value which is decremented down to zero and tested or as a target address in a similar way to the link register.

This is a special purpose register accessed from the user level by the 'move to and move from special purpose' instructions.

Chapter 2
The MPC601
programming model

The MPC601 programming model is best described as being a transitional model between the RS/6000 POWER architecture and the PowerPC. At the user level, it is very similar to the PowerPC 32 bit model, except that some POWER features have been retained — the MQ and RTC registers. If these registers are not used, the MPC601 is compatible with the PowerPC architecture. The MPC601 specific instructions have been included with the PowerPC instruction descriptions.

The supervisor model is also very similar to a full PowerPC implementation but there are some differences any supervisor program must take into account. The overall model is shown in the diagram.

MPC601 user level differences

The MPC601 user programming model is essentially a 32 bit implementation of the PowerPC model. However, there are some important differences in its implementation and that specified by the architecture.

There is a slight difference in behaviour concerning the r0 register. The PowerPC architecture defines load with address update instructions where the address register is r0 and the destination register is the same as the address register as invalid forms and their execution should cause an exception. However, the MPC601 will execute them, but suppresses any address updating and puts an undefined value into the CR0 register if the record option is used. It is best not to use these instruction forms.

MQ register

The MQ register is a hangover from the POWER architecture. It is used to hold the higher 32 bits of a 32 bit by 32 bit multiplication or the remainder from a division operation. It is also involved with some shift instructions. The MQ register allows the MPC601 to execute these POWER instructions and improves compatibility with this architecture. However, to remain PowerPC compatible, these instructions and the MQ register should not be used. The MPC601 specific instructions have been included in Chapter 6.

This is a special purpose register and is accessed from the user level by the 'move to and move from special purpose' instructions.

RTC registers

The MPC601 has two real time clock registers, driven from an external clock, which provide a high resolution method of measuring time. The lower register, RTCL, is a 32 bit register with bits 0, 1 and 25 to 31 set to zero to create a 23 bit counter. This is incremented by the RTC clock signal. When it overflows it wraps round back to zero and the upper register, which acts as a 32 bit counter, RTCU, is incremented by one.

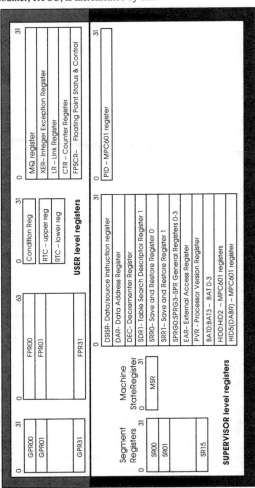

USER level registers

MQ register	
XER – Integer Exception Register	
LR – Link Register	
CTR – Counter Register	
FPSCR – Floating Point Status & Control	

Condition Reg
RTC – upper reg
RTC – lower reg

GPR00
GPR01
GPR31

FPR00
FPR01
FPR31

PID – MPC601 register

SUPERVISOR level registers

DSISR – Data/source instruction register
DAR – Data Address Register
DEC – Decrementer Register
SDR1 – Table Search Descriptor Register 1
SRR0 – Save and Restore Register 0
SRR1 – Save and Restore Register 1
SPRG0:SPRG3 – SPR General Registers 0-3
EAR – External Access Register
PVR – Processor Version Register
BAT0:BAT3 – BAT 0-3
HID0:HID2 – MPC601 registers
HID5(DABR) – MPC601 register

Machine State Register
MSR

Segment Registers
SR00
SR01
SR15

The MPC601 programming model

Care must be accessed when reading these registers to ensure that neither has been updated. The RTC clock rate should be set so that at least 10 add immediate instructions can be executed in a clock period. The recommended process for reading these registers is to read RTCU first, followed by the RTCL and then to read RTCU again and compare it with the first reading. If the values are not the same, the operation is repeated. If only the contents of one of the registers is needed, the register can be simply read.

These are special purpose registers accessed from the user level by the 'move to and move from special purpose' instructions.

These registers do not form any part of the PowerPC architecture and will not be implemented on future PowerPC processors. They are specific to the MPC601 processor. To maintain PowerPC compatibility, these registers should not be used. The PowerPC equivalent is the TB supervisor register, which provides a simple free running timebase.

FPSCR differences

The MPC601 does not implement the VXSOFT and VXSQRT flags in the FPSCR register.

MPC601 supervisor level registers

Machine state register

The most important of the MPC601 supervisor registers is the machine state register due to its control function.

000000000000000	EE	PR	FP	ME	FE0	SE	0	FE1	0	EP	IT	DT	00	0	0
0	15	16	17	18	19	20	21	22	23	24	25	26	27	28 29	30	31

Reserved

Machine state register (MSR) for the MPC601

Flag	Description
EE	External interrupt enable. If cleared to 0, decrementer or external interrupt exceptions are delayed. If set to 1, they can be immediately taken.
PR	Privilege level. If set to 1, the processor can only execute user instructions and is in user mode. If cleared to zero, it can execute both user and supervisor instructions — supervisor mode.
FP	Floating point enable. If cleared to zero, floating point instructions cannot be dispatched, but the registers can still be accessed and exceptions can still occur.
ME	Machine check enable. If cleared to zero, machine check exceptions are disabled. If set to one, they are enabled.
FE0, FE1	These bits control the floating point exception mode as follows:

FE0	FE1	Mode
0	0	Exceptions disabled.
0	1	Imprecise non-recoverable.
1	0	Imprecise recoverable.
1	1	Precise mode

| SE | Single step mode. If cleared to zero, the processor executes normally. If set to one, it will single step through code in correct |

	program order and take a single step exception after every completed instruction.
EP	Exception prefix. This effectively relocates the vector table. If set to one, it is located at $FFFxxxxx. If cleared to zero, it is located at $000xxxxx.
IT	Enable instruction address translation. If cleared to zero, translation is disabled. If set to one, translation is enabled.
DT	Enable data address translation. If cleared to zero, translation is disabled. If set to one, translation is enabled.

MSR flag descriptions for the MPC601

Decrementer register

DEC

0   31

MPC601 decrementer register — DEC

The decrementer register is a 32 bit counter driven from the main processor clock (or subdivision) that generates an exception as it passes through zero (i.e. bit zero changes from 0 to 1). This exception can be caused by normal counting or by writing the contents of any general purpose register into it. The MPC601 is unique in that this supervisor register can be accessed by the user by specifying special purpose register 6. This is to maintain compatibility with the POWER architecture.

The resultant exception can be disabled by clearing the EE bit in the MSR.

External access register

This register is used in conjunction with the eciwx and ecowx instructions which control the special 'external access' mode. If the E bit is cleared, the execution of these two instructions causes a data access exception which allows the supervisor to vet any such access. The RID bits specify the resource ID which is used to identify the external control device. All other bits are set to zero. Please note that the MPC601 only uses bits 28 to 31 for the RID. Other PowerPC processors use bits 26 to 31.

E	EAR	RID

1   28   31

MPC601 external access control register

Special purpose registers

SPRG0
SPRG1
SPRG2
SPRG3

0   31

MPC601 special purpose general registers

These are additional registers that the supervisor can access for its own use.

Processor version register

This register contains the PowerPC processor version number and the particular revision level. The revision level refers to an engineering change level and although it may have some reference to mask revisions, this is not always the case. The initial revision value is $0000. For the MPC603, the version number is $0003 and for the MPC601 it is $0001. The version number allows software to adapt to a particular processor, thus allowing a single software release containing supervisor code to support multiple platforms

VERSION	REVISION
0 15	16 31

MPC601 processor version register

Segment register

There are sixteen segment registers to cover the sixteen 256 Mbyte segments within the 32 bit logical memory map. These registers are in the supervisor programming model and are used as part of the memory management system. Two formats are used, depending on whether bit 0 is set or not. This T bit indicates whether the segment is normal memory or an I/O controller interface.

With T=0, the register refers to a normal memory segment. The virtual segment ID (VSID) is stored in the upper 24 bits while the supervisor and user state storage keys are located in bits 1 and 2 respectively. These key bits are used with the protection bits for both block and page address translation. The use of these keys and their registers are fully described in the chapter on memory management.

T	KS	KU	00000	VSID
0	1	2	3 8	31

▨ Reserved

Bit	Name	Descripton
0	T	T=0 selects this format (paging).
1	KS	Supervisor state storage Key.
2	KU	User state storage key.
8:31	VSID	Virtual Segment ID.

Segment register format with T = 0 (MCP601)

T	KS	KU	BUID	Controller specific
0	1	2	3 12	31

Bit	Name	Description
0	T	T=1 selects this format (I/O Controller interface).
1	KS	Supervisor state storage key
2	KU	User state storage key
3:11	BUID	Bus Unit ID.
12:31		Device dependent data for I/O controller

Segment register with T = 1 (MPC601)

Block address translation registers

Each block address translation (BAT) register pair consists of two 32 bit registers, where the upper register stores the translation and protection information and the lower contains the block definition itself. The full description of these registers is described in the chapter on memory management.

BRPN		00000000	KS	KU	WIMG	0	PP
0	14 15	22 23	24	25	28 29	30	31

Bit	Name	Description
0:14	BRPN	Block Real Page Number
23	KS	Supervisor state storage key
24	KU	User (problem) state storage key
25:28	WIMG	Storage Access controls
30:31	PP	Protection bits for BAT area

▢ Reserved

Upper BAT register(MPC601)

BEPI		00000		BL	V
0	14 15	19 20		30	31

Bit	Name	Description
0:14	BEPI	Block Effective Page Index
20:30	BL	Block Length
31	V	BAT pair valid if V = 1

▢ Reserved

Lower bat register(MPC601)

The SDR1 register

HTABORG		0000000	HTABMASK
0	15 16	22 23	31

▢ Reserved

The SDR1 register format

The SDR1 register is used to control the hashing function used in the memory management system It has two fields: HTABORG and HTABMASK. HTABORG is the upper 16 bits of the start or origin of the page table entry groups (PTEGs). HTABMASK is a field which determines how many bits from the hashing function are used as the index into the collection of PTEGs located by the address in HTABORG. The use of the hashing function is fully described in the chapter on memory management.

Exception processing registers

These registers are used during exception processing. SRR0 is used to hold the instruction address to be used after a return from interrupt instruction signals the restart of normal processing. It is also used to hold the return address with the system call instruction. SRR1 is used to store relevant exception information: bits 0 to 15 are loaded with descriptive information while bits 16 to 31 are a copy of bits 16-31 of the MSR. DSISR and DAR are used to provide more information on the exception cause. Their settings are described in the chapter on exception processing.

DAE/Source Instruction Service Register
DAR- Data Address Register
SRR0 - Save and Restore Register 0
SRR1 - Save and Restore Register 1

0   31

MPC603 special purpose registers for exception processing

HID 0 register

This register controls the checkstop checking that the MPC601 uses to confirm its integrity. The signals fall into two types: those that enable certain checks and those that indicate an error condition, if set.

Bit	Name	Description
0	CE	Master checkstop enable.
1	S	Microcode checkstop detected if set.
2	M	Double machine check detected if set.
3	TD	Multiple TLB hit checkstop if set.
4	CD	Multiple cache hit checkstop if set.
5	SH	Sequencer time-out checkstop if set.
6	DT	Dispatch time-out checkstop if set.
7	SA	Bus address parity error if set.
8	BD	Bus data parity error if set.
9	CP	Cache parity error if set.
10	IU	Invalid microcode instruction if set.
11	PP	I/O controller interface access protocol error if set.
12-14	—	Reserved
15	ES	Enable microcode checkstop. Enabled by hard reset. Enabled if set.
16	EM	Enable machine check checkstop. Dis-abled by hard reset. Enabled if set.
17	ETD	Enable TLB checkstop. Disabled by hard reset. Enabled if set.
18	ECD	Enable cache checkstop. Disabled by hard reset, Enabled if set.
19	ESH	Enable sequencer time-out checkstop. Disabled by hard reset. Enabled if set.
20	EDT	Enable dispatch time-out checkstop. Disabled by hard reset. Enabled if set.
21	EBA	Enable bus address parity checkstop. Disabled by hard reset. Enabled if set.
22	EBD	Enable bus data parity checkstop. Dis-abled by hard reset. Enabled if set.
23	ECP	Enable cache parity checkstop. Dis-abled by hard reset. Enabled if set.
24	EIU	Enable for invalid ucode instruction checkstop. Enabled by hard reset and if set to one.
25	EPP	Enable for I/O controller interface access protocol checkstop. Disabled by hard reset. Enabled if set,
26	DRF	0 Optional reload of alternate sector on instruction fetch miss is enabled.
		1 Optional reload of alternate sector on instruction fetch miss is disabled.

27	DRL	0 Optional reload of alternate sector on load/store miss is enabled. 1 Optional reload of alternate sector on load/store miss is enabled.
28	LM	0 Big-endian mode is enabled. 1 Little-endian mode is enabled.
29	PAR	0 Precharge of the ARTRY* and SHD* signals is enabled. 1 Precharge of the ARTRY* and SHD* signals is disabled.
30	EMC	0 No error detected in main cache during initialisation. 1 Error detected in main cache during initialisation.
31	EHP	0 The HP_SNP_REQ* signal is disabled. 1 The HP_SNP_REQ* signal is enabled.

The HID0 checkstop register definitions

HID 1 register

This is the debug mode register. It controls the debug functions within the MPC601.

0	M	0000	RM	0000000	MISC
0 1	3 4	7 8	9 10	16 17	31

□ Reserved

The HID1 register

Field	Code	Description
M	000	Normal run mode
	001	Undefined — do not use!
	010	Limited instruction address compare.
	011	Undefined — do not use!
	100	Single instruction step
	101	Undefined — do not use!
	110	Full instruction address compare.
	111	Full data address compare.
RM	00	Hard stop.
	01	Soft stop - waits for system activity to stop.
	10	Trap to run mode exception vector '0 2000'.
	11	Reserved — do not use!
MISC	—	Bit 17 when high disables the broadcast of the tlbie instruction. All other bits in this field are reserved and should not be used.

The HID1 register bit descriptions

HID 2 register — IABR

The instruction address breakpoint register (IABR) contains the 32 bit address to be used for an instruction address breakpoint. It is used in conjunction with the HID1 register to enable these types of breakpoints. Although the register is 32 bits wide, bits 30 and 31 are set to zero and are reserved. These bits are not needed to identify the instruction address as it always on a 32 bit boundary. If a breakpoint is triggered, this generates an exception which is then processed by a specific exception handler.

HID 5 register — DABR

The data address breakpoint register (DABR) contains the 32 bit address to be used for a data address breakpoint. It is used in conjunction with the HID1 register to enable these types of breakpoints. Although the register is 32 bits wide, bits 29 through to 31 are not part of the address and are used for other purposes. Because of this, the address comparison is performed using the double word address and the lower three bits are effectively ignored.

Bit 29 is set to zero and is reserved. Bits 30 and 31 form the SA field, which controls the memory access that can trigger a breakpoint exception.

Code	Description
00	Breakpoint disabled.
01	Breakpoint load accesses only.
10	Breakpoint store accesses only.
11	Breakpoint both store and load accesses.

SA field description

HID 15 register — PIR

The processor identification register (PIR) contains a four-bit identification tag in bits 29 to 31; all other bits are set to zero and reserved. The tag is used to identify the processor in multi-processor systems when accessing I/O controller interface segments.

Chapter 3

The MPC603 programming model

The MPC603 programming model is a true PowerPC 32 bit implementation. It has the normal PowerPC user programming model and does not have the RTC and MQ registers that the MPC601 has.

USER level registers

GPR00
GPR01
GPR31

FPR00
FPR01
FPR31

Condition Reg

XER – Integer Exception Register
LR – Link Register
CTR – Counter Register
FPSCR– Floating Point Status & Control

SUPERVISOR level registers

Segment Registers
SR00
SR01
SR15

Machine State Register
MSR

DEC– Decrementer Register
SDR1 – Table Search Descriptor Register 1
DSISR– Data/source instruction register
DAR– Data Address Register
SRR0– Save and Restore Register 0
SRR1– Save and Restore Register 1
SPRG0:SPRG3–SPR General Registers 0-3
EAR – External Access Register
PVR – Processor Version Register
IBAT0:IBAT3 – Instruction BAT 0-3
DBAT0:DBAT3 – Data BAT 0-3

TB – time base
TBU – time base upper
DMISS – data TLB miss address
IMISS – instruction TLB miss address
DCMP – data TLB compare value
ICMP – instruction TLB compare value
HASH1 – first PTEG hashed address
HASH2 – second PTEG hashed address
RPA – real page address
HID0 – hardware implementation register
HID1 – instruction address breakpoint (IABR)

MPC603 programming model

The supervisor model is also different from that of the MPC601, although there is some commonality. The reason for this is partially due to the different MMU and cache memory organisation and the fact that the MPC603 performs external software table walks through software, whereas the MPC601 does them through hardware without software intervention. To allow software to perform this function, the MPC603 has some additional registers which provide it with the address and hashing information it needs.

It should be remembered that while the PowerPC architecture does define some common supervisor level registers, the supervisor programming model is implementation dependent. It therefore varies from processor to processor. This does not affect user program compatibility as these programs have the same model across the PowerPC processor family. However, it does affect operating system software, which may require modification to move an operating system environment from one processor to another.

The MPC603 user programming model

As shown overleaf, this is consistent with the PowerPC user model and consists of 32 general purpose registers, 32 floating point registers, floating point status and control register, condition register, link and integer exception registers.

The time base registers have been shown as part of the supervisor model. This is not strictly accurate because the two registers can be read by the user without supervisor status or permission. Writing to these registers however is a supervisor level action and cannot be performed by the user.

MPC603 supervisor level registers

Although some of the MPC603 supervisor registers are similar to those of the MPC601, it should not be assumed that the supervisor programming model is the same — it is not!

Machine state register

000000000000000	EE	PR	FP	ME	FEO	SE	BE	FE1	0	EP	IT	DT	00	RE	LE
0	15	16	17	18	19	20	21	22	23	24	25	26	27	28	29 30	31

▨ Reserved

Machine state register (MSR) for the MPC603

Flag	Description
EE	External interrupt enable. If cleared to 0, decrementer or external interrupt exceptions are delayed. If set to 1, they can be immediately taken.
PR	Privilege level. If set to 1, the processor can only execute user instructions and is in user mode. If cleared to zero, it can execute both user and supervisor instructions — supervisor mode.

FP	Floating point enable. If cleared to zero, floating point instructions cannot be dispatched, but the registers can still be accessed and exceptions can still occur.
ME	Machine check enable. If cleared to zero, machine check exceptions are disabled. If set to one, they are enabled.
FE0, FE1	These bits control the floating point exception mode as follows:

FE0	FE1	Mode
0	0	Exceptions disabled.
0	1	Imprecise non-recoverable.
1	0	Imprecise recoverable.
1	1	Precise mode

SE	Single step mode. If cleared to zero, the processor executes normally. If set to one, it will single step through code in correct program order and take a single step exception after every completed instruction.
BE	Branch trace enable. If cleraed to zero, branch instructions are executed normally. If set to 1, a trace exception is generated after each complete branch to allow program tracing.
EP	Exception prefix. This effectively relocates the vector table. If set to one, it is located at $FFFxxxxx. If cleared to zero, it is located at $000xxxxx.
IT	Enable instruction address translation. If cleared to zero, translation is disabled. If set to one, translation is enabled.
DT	Enable data address translation. If cleared to zero, translation is disabled. If set to one, translation is enabled.
RE	Recoverable exception. If cleared to zero, the interrupt is not recoverable. If set to one, it is recoverable. This is used by some interrupt handlers to indicate whether the processor should continue processing or shut down.
LE	Little endian mode. If cleared to zero, the processor runs in big endian mode. If set to 1, it runs in little endian mode. This change requires a speciacode sequence using an rfi instruction.

MSR flag descriptions for the MPC603

There are two bits specified in the PowerPC architecture which are not implemented on either the MPC601 or MPC603. The first is bit 31, which controls the endian organisation of the external memory bus, and the second is bit 22, which controls branch trace exceptions.

Decrementer register

DEC
0   31

MPC603 decrementer register — DEC

The decrementer register is a 32 bit counter driven from the main processor clock (or subdivision) which generates an exception as it passes through zero (i.e. bit zero changes from 0 to 1). This exception can be caused by normal counting or by writing the contents of any general purpose register into it.

The resultant exception can be disabled by clearing the EE bit in the MSR.

Processor version register

This register contains the PowerPC processor version number and the particular revision level. The revision level refers to an engineering change level and, although it may have some reference to mask revisions, this is not always the case. The initial revision value is $0000. For the MPC603, the version number is $0003 and for the MPC601 it is $0001. The version number allows software to adapt to a particular processor, thus allowing a single software release containing supervisor code to support multiple platforms.

VERSION	REVISION
0 15	16 31

MPC603 processor version register

External access register

This register is used in conjunction with the eciwx and ecowx instructions which control the special 'external access' mode. If the E bit is cleared, the execution of these two instructions causes a data access exception, which allows the supervisor to vet any such access. The RID bits specify the resource ID which is used to identify the external control device. All other bits are set to zero. Please note that the MPC601 uses a different number of bits for the RID.

E	RESERVED	RID
0 1	25 26	31

MPC603 external access control register

Special purpose registers

These are additional registers that the supervisor can access for its own use.

SPRG0
SPRG1
SPRG2
SPRG3
0   31

MPC603 special purpose general registers

Time base registers

These are two 32 bit registers used to provide a time base function. Although split across the two registers, they form a 64 bit counter used to provide a time reference for software. With the MPC603, the lower 32 bits are in the TB register while the TBU contains the upper bits. These registers must be initialised in software, if required, and are driven from the processor clock or subdivision. They do not generate an interrupt or exception. In this respect, they are a simpler form of the MPC601 RTC registers.

The recommended process for reading these registers is to read TBU first, followed by the TB(some documentation and the MPC604 use the name TBL, and use TB to refer to both TBL and TBU as a pair) and to read TBU again and compare it with the first reading. If the values are not the same, the operation is repeated. If only the contents of one of the registers is needed, the register can be simply read.

They can be used to timestamp events within the system and to track how long and when a task was last executed. This information can then be used as the basis of a scheduling algorithm which decides which tasks get priority and are executed first.

As previously mentioned, these registers can be read but not written to by the user while the supervisor has full read and write access.

Exception processing registers

These registers are used during exception processing. SRR0 is used to hold the instruction address to be used after a return from interrupt instruction signals the restart of normal processing. It is also used to hold the return address with the system call instruction. SRR1 is used to store relevant exception information: bits 0 to 15 are loaded with descriptive information while bits 16 to 31 are a copy of bits 16-31 of the MSR. DSISR and DAR are used to provide more information on the exception cause. Their settings are described in the chapter on exception processing.

DAE/Source Instruction Service Register
DAR- Data Address Register
SRR0 - Save and Restore Register 0
SRR1 - Save and Restore Register 1

0   31

MPC603 special purpose registers for exception processing

Segment register

There are sixteen segment registers to cover the sixteen 256 Mbyte segments within the 32 bit logical memory map. These registers are in the supervisor programming model and are used as part of the memory management system. Two formats are used, depending on whether bit 0 is set or not. This T bit indicates whether the segment is normal memory or an I/O controller interface.

With T=0, the register refers to a normal memory segment. The virtual segment ID (VSID) is stored in the upper 24 bits while the supervisor and user state storage keys are located in bits 1 and 2 respectively. These key bits are used with the protection bits for both block and page address translation.

T	KS	KU	00000	VSID
0	1	2 3	8	31

▨ Reserved

Bit	Name	Descripton
0	T	T=0 selects this format (paging).
1	KS	Supervisor state storage key.
2	KU	User state storage key.
8:31	VSID	Virtual Segment ID.

Segment register format with T = 0 (MCP603)

If the T bit is set to 1, this is considered an access to the I/O controller interface and the coding of the bits is different.

T	KS	KU	BUID	Controller specific
0	1	2	3 11	12	31

Bit	Name	Description
0	T	T=1 selects this format (I/O Controller interface).
1	KS	Supervisor state storage key
2	KU	User state storage key
3:11	BUID	Bus Unit ID.
12:31		Device dependent data for I/O controller

Segment register with T = 1 (MPC603)

The SDR1 register

HTABORG	0000000	HTABMASK
0 15	16 22	23 31

▨ Reserved

The SDR1 register format

The SDR1 register is used to control the hashing function used in the memory management system It has two fields: HTABORG and HTABMASK. HTABORG is the upper 16 bits of the start or origin of the page table entry groups (PTEGs). HTABMASK is a field which determines how many bits from the hashing function are used as the index into the collection of PTEGs located by the address in HTABORG. The use of the hashing function is fully described in the chapter on memory management.

Data and instruction block address translation registers

The IBAT and DBAT register pairs both consist of two 32 bit registers, where the upper register stores the translation and protection information and the lower contains the block definition itself.

BEPI		00000		BL		V
0	14 15		19 20		30	31

Bit	Name	Description
0:14	BEPI	Block Effective Page Index
20:30	BL	Block Length
31	V	BAT pair valid if V = 1

□ Reserved

Lower DBAT or IBAT register(MPC603)

BRPN		00000000		KS	KU	WIMG	0	PP
0	14 15		22	23	24	25	28 29	30 31

Bit	Name	Description
0:14	BRPN	Block Real Page Number
23	KS	Supervisor state storage key
24	KU	User (problem) state storage key
25:28	WIMG	Storage Access controls
30:31	PP	Protection bits for BAT area

□ Reserved

Upper DBAT or IBAT register(MPC603)

The SDR1 register

HTABORG		0000000		HTABMASK	
0	15	16	22	23	31

□ Reserved

The SDR1 register format

The SDR1 register is used to control the hashing function used in the memory management system It has two fields: HTABORG and HTABMASK. HTABORG is the upper 16 bits of the start or origin of the page table entry groups (PTEGs). HTABMASK is a field which determines how many bits from the hashing function are used as the index into the collection of PTEGs located by the address in HTABORG.

MPC603 specific registers

The remaining registers are MPC603 specific. The first seven provide information to allow the MPC603 to perform memory management table walks in software.

DMISS - Data TLB miss address
IMISS - Instruction TLB miss address
DCMP - Data TLB compare value
ICMP - Instruction TLB compare value
HASH1 - First PTEG hashed address
HASH2 - Second PTEG hashed address
RPA - Real Page Address
HID0 - Hardware implementation register
HID1 - Instruction address breakpoint (IABR)

0	31

The remaining two registers provide control over the processor's power management and hardware specific functions, including the instruction address breakpoint facility.

These registers are provided for the supervisor to use when handling memory management exceptions and for controlling the processors's machine checking.

HID0 — Hardware implementation register

This register is processor dependent and controls the internal machine checking that is performed within the processor. This register is an extremely dangerous one to manipulate and should be left alone or left to the operating system to handle or manipulate. Many of the bits affect the external bus and hardware and changing these settings may cause strange effects.

Bit	Name	Description
0	EMCP	Enable machine check pin.
2	EBA	Enable bus address parity checking.
3	EBD	Enable bus data parity checking.
4	SBCLK	Select bus clock for test clock pin.
5	EICE	Enable ICE outputs for pipeline tracking.
6	ECLK	Enable external test clock pin.
7	PAR	Disable precharge of shared signals.
8	DOZE	Enable doze mode(PLL, time base and bus snooping active).
9	NAP	Enable nap mode (PLL and time base active).
10	SLEEP	Enable sleep mode.
11	DPM	Enable dynamic power management.
12	RISEG	Reserved for test.
15	NHR	Reserved.
16	ICE	Instruction cache enable.
17	DCE	Data cache enable.
18	ILOCK	Instruction cache lock.
19	DLOCK	Datacache lock.
20	ICFI	Instruction cache flash invalidate.
21	DCI	Data cache flash invalidate.
27	FBIOB	Force branch indirect on bus.
31	NOOPTI	No-op touch instructions.

All other bits are not used.

HID2 — Instruction address breakpoint register

This register is used to contain the address to be used for controlling an instruction breakpoint. It should be remembered that the instruction breakpoint mechanism does require an appropriate exception handler to process the exception that is generated by the breakpoint. The breakpoint is enabled through the use of bit 30.

Chapter 4

The MPC604 programming model

The MPC604 programming model is a true PowerPC 32 bit implementation. It has the normal PowerPC user programming model and again like the MPC603, does not have the RTC and MQ registers that the MPC601 has.

USER level registers

XER	Integer Exception Register
LR	Link Register
CTR	Counter Register
FPSCR	Floating Point Status & Control
TBL	time base
TBU	time base upper
HID0	hardware implementction register
PMC1	Performance monitor counter
PMC2	Performance monitor counter
MMCR0	Monitor mode control register
SDA	Sampled data address
SIA	Sampled instruction address
PIR	processor identification register
DABR	data address breakpoint register
IABR	Instruction address breakpoint register

Condition Reg

FPR00, FPR01 ... FPR31

GPR00, GPR01 ... GPR31

SUPERVISOR level registers

DEC	Decrementer Register
SDR1	Table Search Descriptor Register 1
DSISR	Data/source instruction register
DAR	Data Address Register
SRR0	Save and Restore Register 0
SRR1	Save and Restore Register 1
SPRG0:SPRG3	SPR General Registers 0-3
EAR	External Access Register
PVR	Processor Version Register
IBAT0:IBAT3	Instruction BAT 0-3
DBAT0:DBAT3	Data BAT 0-3

Machine State Register — MSR

Segment Registers — SR00, SR01 ... SR15

MPC604 programming model

The supervisor model is also different from that of the MPC601, although there is some commonality. The reason for this is partially due to the different MMU and cache memory organisation and the fact that the MPC604 performs external software table walks through hardware, whereas the MPC603 does them through software intervention. As a result the registers used by the MPC603 to enable the software table walk have gone.

The MPC604 does have some new registers that allow its performance to be monitored. The five new registers allow software to non-obtrusively monitor the processor and gain information concerning different aspects of its operation.

Again, it should be remembered that while the PowerPC architecture does define some common supervisor level registers, the supervisor programming model is implementation dependent. It therefore varies from processor to processor. This does not affect user program compatibility as these programs have the same model across the PowerPC processor family. However, it does affect operating system software, which may require modification to move an operating system environment from one processor to another.

The MPC604 user programming model

As shown, this is consistent with the PowerPC user model and consists of 32 general purpose registers, 32 floating point registers, floating point status and control register, condition register, link and integer exception registers.

MPC604 supervisor level registers

Although some of the MPC604 supervisor registers are similar to those of the MPC603, it should not be assumed that the supervisor programming model is the same — it is not! For example, all the registers that were provided by the MPC603 to perform software table walking are missing because the MPC604 uses a hardware based technique instead.

Machine state register

000000000000000	EE	PR	FP	ME	FE0	SE	BE	FE1	0	EP	IT	DT	0	PM	RE	LE
0	15	16	17	18	19	20	21	22	23	24	25	26	27	28	29	30	31

▨ Reserved

Machine state register (MSR) for the MPC604

The MPC604 MSR differs from that of the MCP603 through the definition of bit 29 as the PM bit. This is normally reserved.

Flag	Description
EE	External interrupt enable. If cleared to 0, decrementer or external interrupt exceptions are delayed. If set to 1, they can be immediately taken.
PR	Privilege level. If set to 1, the processor can only execute user instructions and is in user mode. If cleared to zero, it can execute both user and supervisor instructions — supervisor mode.
FP	Floating point enable. If cleared to zero, floating point instructions cannot be

	dispatched, but the registers can still be accessed and exceptions can still occur.
ME	Machine check enable. If cleared to zero, machine check exceptions are disabled. If set to one, they are enabled.
FE0, FE1	These bits control the floating point exception mode as follows:

FE0	FE1	Mode
0	0	Exceptions disabled.
0	1	Imprecise non-recoverable.
1	0	Imprecise recoverable.
1	1	Precise mode

SE	Single step mode. If cleared to zero, the processor executes normally. If set to one, it will single step through code in correct program order and take a single step exception after every completed instruction.
BE	Branch trace enable. If cleraed to zero, branch instructions are executed normally. If set to 1, a trace exception is generated after each complete branch to allow program tracing.
EP	Exception prefix. This effectively relocates the vector table. If set to one, it is located at $FFFxxxxx. If cleared to zero, it is located at $000xxxxx.
IT	Enable instruction address translation. If cleared to zero, translation is disabled. If set to one, translation is enabled.
DT	Enable data address translation. If cleared to zero, translation is disabled. If set to one, translation is enabled.
PM	This bit allows the currently executing software — the process — to indicate if the processor monitoring should be enabled or not. If it is cleared to 0, then the monitoring is not enabled. If it is set to 1, it is.
RE	Recoverable exception. If cleared to zero, the interrupt is not recoverable. If set to one, it is recoverable. This is used by some interrupt handlers to indicate whether the processor should continue processing or shut down.
LE	Little endian mode. If cleared to zero, the processor runs in big endian mode. If set to 1, it runs in little endian mode. This change requires a speciacode sequence using an rfi instruction.

MSR flag descriptions for the MPC604

The recommended method for changing the endian mode of the processor uses the SRR0 and SRR1 registers and can only be performed by the supervisor. A copy of the MSR is made and the LE bit changed accordingly. The modified copy of the MSR is then copied into SRR1 and the starting address of the new code is placed into SRR0. An rfi is then executed to switch the processor.

Decrementer register

DEC
0   31

MPC604 decrementer register — DEC

The decrementer register is a 32 bit counter driven from the main processor clock (or subdivision) which generates an exception as it passes through zero (i.e. bit zero changes from 0 to 1). This exception can be caused by normal counting or by writing the contents of any general purpose register into it. The resultant exception can be disabled by clearing the EE bit in the MSR.

Processor version register

This register contains the PowerPC processor version number and the particular revision level. The revision level refers to an engineering change level and, although it may have some reference to mask revisions, this is not always the case. The initial revision value is $0000. For the MPC604, the version number is $0004. The version number allows software to adapt to a particular processor, thus allowing a single software release containing supervisor code to support multiple platforms. There may be multiple revisions number and care should be exercised when using this register to check the processor type.

VERSION	REVISION
0 15	16 31

MPC604 processor version register

External access register

This register is used in conjunction with the eciwx and ecowx instructions which control the special 'external access' mode. If the E bit is cleared, the execution of these two instructions causes a data access exception, which allows the supervisor to vet any such access. The RID bits specify the resource ID which is used to identify the external control device. All other bits are set to zero. Unlike the MPC601, the MPC604 allows the use of bits 26 to 31 for the RID as specified in the PowerPC architecture.

E	RESERVED	RID
0 1	25 26	31

MPC604 external access control register

Special purpose registers

These are additional registers that the supervisor can access for its own use.

SPRG0
SPRG1
SPRG2
SPRG3
0   31

MPC604 special purpose general registers

Time base registers

These are two 32 bit registers used to provide a time base function. Although split across the two registers, they form a 64 bit counter used to provide a time reference for software. With the MPC604, the lower 32 bits are in the TBL register while the TBU contains the upper bits. These registers must be initialised in software, if required, and are driven from the processor clock or subdivision. They do not generate an interrupt or exception. In this respect, they are a simpler form of the MPC601 RTC registers.

Although they are the same as the time base registers specified in the MPC603 implementation, the documentation is not consistent on how the TBL register is named and sometime uses TB instead. The name TB in turn is often used to refer to both registers together,

The recommended process for reading these registers is to read TBU first, followed by the TBL and to read TBU again and compare it with the first reading. If the values are not the same, the operation is repeated. If only the contents of one of the registers is needed, the register can be simply read.

They can be used to timestamp events within the system and to track how long and when a task was last executed. This information can be used as the basis of a scheduling algorithm which decides which tasks get priority and are executed first.

To facilitate this, they can be read from the user programming environment. However, writing to them is a supervisor level operation.

Exception processing registers

These registers are used during exception processing. SRR0 is used to hold the instruction address to be used after a return from interrupt instruction signals the restart of normal processing. It is also used to hold the return address with the system call instruction. SRR1 is used to store relevant exception information: bits 0 to 15 are loaded with descriptive information while bits 16 to 31 are a copy of bits 16-31 of the MSR. DSISR and DAR are used to provide more information on the exception cause. Their settings are described in the Chapter 5 on exception processing.

DAE/Source Instruction Service Register
DAR- Data Address Register
SRR0 - Save and Restore Register 0
SRR1 - Save and Restore Register 1

0   31

MPC604 special purpose registers for exception processing

Segment register

There are sixteen segment registers to cover the sixteen 256 Mbyte segments within the 32 bit logical memory map. These registers are in the supervisor programming model and are used as part of the memory management system. Two formats are used, depending on whether bit 0 is set or not. This T bit indicates whether the segment is normal memory or an I/O controller interface.

With T=0, the register refers to a normal memory segment. The virtual segment ID (VSID) is stored in the upper 24 bits while the supervisor and user state storage keys are located in bits 1 and 2 respectively. These key bits are used with the protection bits for both block and page address translation. The use of these keys and their registers are fully described in the chapter on memory management.

T	KS	KU	00000	VSID
0	1	2	3	8	31

▨ Reserved

Bit	Name	Descripton
0	T	T=0 selects this format (paging).
1	KS	Supervisor state storage key.
2	KU	User state storage key.
8:31	VSID	Virtual Segment ID.

Segment register format with T = 0 (MCP604)

If the T bit is set to 1, this is considered an access to the I/O controller interface and the coding of the bits is different.

T	KS	KU	BUID	Controller specific
0	1	2	3	12	31

Bit	Name	Description
0	T	T=1 selects this format (I/O Controller interface).
1	KS	Supervisor state storage key
2	KU	User state storage key
3:11	BUID	Bus Unit ID.
12:31		Device dependent data for I/O controller

Segment register with T = 1 (MPC604)

The SDR1 register

HTABORG	0000000	HTABMASK
0	15 16	22 23	31

▨ Reserved

The SDR1 register format

The SDR1 register is used to control the hashing function used in the memory management system It has two fields: HTABORG and HTABMASK. HTABORG is the upper 16 bits of the start or origin of the page table entry groups (PTEGs). HTABMASK is a field which determines how many bits from the hashing function are used as the index into the collection of PTEGs located by the address in HTABORG. The use of the hashing function is fully described in the chapter on memory management.

Data and instruction block address translation registers

The IBAT and DBAT register pairs both consist of two 32 bit registers, where the upper register stores the translation and protection information and the lower contains the block definition itself. The full description of these registers is described in the chapter on memory management.

BRPN		00000000	KS	KU	WIMG	0	PP
0	14 15	22 23	24	25	28 29	30	31

Bit	Name	Description
0:14	BRPN	Block Real Page Number
23	KS	Supervisor state storage key
24	KU	User (problem) state storage key
25:28	WIMG	Storage Access controls
30:31	PP	Protection bits for BAT area

☐ Reserved

Upper DBAT or IBAT register(MPC604)

BEPI		00000	BL	V
0	14 15	19 20	30	31

Bit	Name	Description
0:14	BEPI	Block Effective Page Index
20:30	BL	Block Length
31	V	BAT pair valid if V = 1

☐ Reserved

Lower DBAT or IBAT register(MPC604)

The SDR1 register

HTABORG		0000000	HTABMASK
0	15 16	22 23	31

☐ Reserved

The SDR1 register format

The SDR1 register is used to control the hashing function used in the memory management system It has two fields: HTABORG and HTABMASK. HTABORG is the upper 16 bits of the start or origin of the page table entry groups (PTEGs). HTABMASK is a field which determines how many bits from the hashing function are used as the index into the collection of PTEGs located by the address in HTABORG. The use of the hashing function is fully described in the chapter on memory management.

MPC604 specific registers

The remaining registers are MPC604 specific. They include the hardware implementation register and the perfoamance monitor registers that allow system performance information to be obtained with little overhead.

HID0 — hardware implementation register

This register is processor dependent and controls the internal machine checking that is performed within the processor. This register is an extremely dangerous one to manipulate and should be left alone or left to the operating system to handle or manipulate

Bit	Description
0	Enable machine check input pin. If cleared to zero, the pin is disabled. If set to 1, it is enabled.
1	Enable cache parity checking. If cleared to zero, the facility is not enabled. If set to 1, it is.
2	Enable machine check on address bus parity error. If cleared to zero, the facility is not enabled. If set to 1, it is.
3	Enable machine check on data bus parity error. If cleared to zero, the facility is not enabled. If set to 1, it is.
7	Disable snoop response high state restore. If this bit is active, then the external bus protocol is modified.
15	Not hard reset. If cleared to zero, software has cleared this bit , indicating that a hard reset had occurred. If set to 1, a hard reset has not taken place.
16	Instruction cache enable. If cleared to zero, the cache is not enabled. If set to 1, it is.
17	Data cache enable. If cleared to zero, the cacheis not enabled. If set to 1, it is.
18	Instruction cache lock. If cleared to zero, the cache is not locked. If set to 1, it is.
19	Data cache lock. If cleared to zero, the cache is not locked. If set to 1, it is.
20	Instruction cache invalidate all. If cleared to zero, the cache is not invalidated (normal operation). If set to 1, it is.
21	Data cache invalidate all. If cleared to zero, the cache is not invalidated (normal operation). If set to 1, it is.
24	MPC604 serial mode. If cleared to zero, the MPC604 will execute instructions serially and the internal parallelism is disabled. If set to 1, the processor operates normally.
29	Branch history table. If cleared to zero, the static branch prediction is used (the history table is disabled and is the power on default state). If set to 1, the branch history table is enabled.

IABR — Instruction address breakpoint register

This register controls the instrcution breakpoint facility.

Address	BE	TE
0	29 30	31

The IABR register

Flag	Description
BE	Breakpoint enable. If cleared to zero, breakpoint checking is disabled. If set to 1, it is enabled.
TE	Translation enabled. This bit is compared with the instruction translation bit of the MSR and an instruction breakpoint is only signlaed if both bits match.

PIR — Processor identification register

This register is used to provide the processor with a unique identity for use within multiprocessor designs. BY supplying a PID value for each processor in the system, the operating system can uniquely identify each processor and can further use this number to identify system resources.

000000000000000000000000000000		PID
0	27 28	31

▨ Reserved

The PIR register

MMCR0 — Monitor mode control register 0

This register is used control the performance monitoring facilities offered by the MPC604. Two count registers are used (PMC1 and PMC2) are decremented as certain events occur as the processor executes its software. If these counters become negative, they trigger an interrupt which allows software an opportunity to get the performance information. Two other registers (SIA and SDA) hold the data and instruction details at the time of the interrupt.

Bit	Name	Description
0	DIS	Disable counting unconditionally. If cleared to zero, the values of the PMCn counters can be updated by hardware.
1	DP	Disable counting while in supervisor mode. If cleared to zero, the PMCn counters can be changed by hardware irrespective of the user/supervisor mode. If set to 1, the updating is disabled while the processor is in supervisor mode.
2	DU	Disable counting while in user mode. If cleared to zero, the PMCn counters can be changed by hardware irrespective of the user/supervisor mode. If set to 1, the updating is disabled while the processor is in user mode.
3	DMS	Disable counting while MSR[PM] is set. If cleared to zero, the PMCn counters can be changed by hardware irrespective of the state of MSR[PM] bit. If set to 1 and MSR[PM] is set to 1, the updating is disabled.
4	DMR	Disable counting while MSR(PM) is zero. If cleared to zero, the PMCn counters can be

changed by hardware irrespective of the state of MSR[PM] bit. If set to 1 and MSR[PM] is cleared to 0, the updating is disabled.

5 ENINT Enable performance monitoring interrupt signaling. If this bit is cleared to zero, then no interrupt is generated. If set to one, the interrupt generation is enabled. The bit is automatically cleared when a performance interrupt is signalled. The exception handling routine must set the bit to allow the performance monitoring to continue.

5 DISCOUNT

Disable counting of PMC1 and PMC2 when a performance monitor interrupt is generated. If cleared to zero, performance monitoring will continue during the exception handling to process the performance monitor interrupt. If set to 1, the updating of the PMC1 is stopped and PMC2 will not go negative, thus disabling the performance monitoring.

7–8 RTCSELECT

64-bit time base, bit selection enable

00	Pick bit 63 to count
01	Pick bit 55 to count
10	Pick bit 51 to count
11	Pick bit 47 to count

9 INTONBITTRANS

Generate performance monitor interrupt when RTC bit changes from off to on. If cleared to zero, do not allow interrupt signal if chosen bit transitions. If set to 1, allow the interrupt to be generated.

10–15 THRESHOLD

Threshold value. This value (0 to 63) provides data on level 1 data cache misses.

16 PMC1INTCONTROL

Enable interrupt signaling due to PMC1 counter negative. If cleared to zero, disable PMC1 interrupt signaling when PMC1 counter goes negative. If set to 1, enable it.

17 PMC2INTCONTROL

Enable interrupt signaling due to PMC2 counter negative. If cleared to zero, disable PMC2 interrupt signaling when PMC2 counter goes negative. If set to 1, enable it. This signal overrides bit 6 (DISCOUNT).

18 PMC2COUNTCTL

This can be used to trigger counting of PMC2 after PMC1 has become negative or after a performance monitoring interrupt is signaled. If cleared to zero, enable PMC2 counting. If set to 1, disable PMC2 counting.

19-25 PMC1SELECT

This selects the events that will update PMC1.

Encoding	Description
000 0000	Nothing.
000 0001	Processor cycles.
000 0010	Number of instructions completed.
000 0011	RTCSELECT bit transition.

Encoding	Description
000 0100	Number of instructions dispatched.
000 0101	Icache misses.
000 0110	dtlb misses.
000 0111	Branch predicted incorrectly.
000 1000	Number of reservations requested (LARX is ready for execution).
000 1001	Number of load dcache misses that exceeded the threshold value with lateral L2 intervention.
000 1010	Number of store dcache misses that exceeded the threshold value with lateral L2 intervention.
000 1011	Number of mtspr instructions dispatched.
000 1100	Number of sync instructions.
000 1101	Number of eieio instructions.
000 1110	Number of integer instructions being completed every cycle (no loads or stores).
000 1111	Number of floating-point instructions being completed every cycle (no loads or stores).
001 0000	LSU produced result.
001 0001	SCIU1 produced result.
001 0010	FPU produced result.
001 0011	Instructions dispatched to the LSU.
001 0100	Instructions dispatched to the SCIU1 .
001 0101	Instructions dispatched to the FP unit.
001 0110	Snoop requests received.
001 0111	Number of load dcache misses that exceeded the threshold value without lateral L2 intervention.
001 1000	Number of store dcache misses that exceeded the threshold value without lateral L2 intervention.

26-31 PMC2SELECT

This selects the events that will update PMC2

Encoding	Description
00 0000	Nothing.
00 0001	Processor cycles.
00 0010	Number of instructions completed.
00 0011	RTCSELECT bit transition.
00 0100	Number of instructions dispatched.

Encoding	Description
00 0101	Number of cycles a load miss takes.
00 0110	Data cache misses.
00 0111	Instruction tlb misses.
00 1000	Branches completed.
00 1001	Number of reservations successfully obtained (STCX succeeded).
00 1010	Number of mfspr instructions dispatched.
00 1011	Number of icbi instructions.
00 1100	Number of isync instructions.
00 1101	Branch unit produced result.
00 1110	SCIU0 produced result.
00 1111	MCIU produced result.
01 0000	Instructions dispatched to the branch unit.
01 0001	Instructions dispatched to the SCIU0 .
01 0010	Number of loads completed.
01 0011	Instructions dispatched to the MCIU.
01 0100	Number of snoop hit occurred.

PMC1 — Performance monitor counter register 1

PMC2 — Performance monitor counter register 2

These two registers contain the count values and are normally decremented when ever an event as defined in the MMCR0 register PMC1SELECT and PMC2SELECT fields takes place. They can be modified by software although care must be exercised to ensure that a negative value is not accidently written to these registers. An interrupt is generated when they go negative and thus an erroneous interrupt can be generated.

SIA — Sampled instruction address register

SDA — Sampled data address register

The SIA register holds the address of the instruction that caused a performance monitor interrupt providing the interrupt was caused by a threshold and not some other reason e.g. software writing a negative value into PMC1 or PMC2. In this case, the address is of the last completed instruction. The SDA will hold the effective address of the operand associated with the instruction referred to by the SIA register. Again, this is only valid if the interrupt was generated by a threshold and not some other reason.

Chapter 5
Exception processing

What is an exception?

With reference to the PowerPC architecture, an exception is a transition from the user state to the supervisor state in response to either an external request or error, or some internal condition that requires servicing. Generating an exception is the only way to move from the user state to the supervisor state. Example exceptions include external interrupts, page faults, memory protection violations and bus errors. In many ways the exception handling is similar to that used with CISC pro-cessors, in that the processor changes to the supervisor state, vectors to an exception handler routine, which investigates the exception and services it before returning control to the original program. This general principle still holds but there are fundamental differences which require careful consideration.

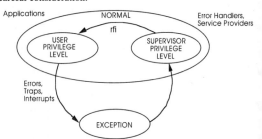

The exception transition model

When an exception is recognised, the address of the instruction to be used by the original program when it restarts and the machine state register (MSR) are stored in the super-visor registers, SRR0 and SRR1. The processor moves into the supervisor state and starts to execute the handler, which resides at the associated vector location in the vector table. The handler can, by examining the DSISR and FPSCR registers, determine the exact cause and rectify the problem or carry out the required function Once completed, the rfi instruction is executed. This restores the MSR and the instruction address from the SRR0 and SRR1 registers and the interrupted program continues.

There are four general types of exception: asynchronous precise or imprecise and synchronous precise and imprecise. Asynchronous and synchronous refer to when the exception is caused: a synchronous exception is one that is synchronised, i.e. caused by the instruction flow. An asynchronous exception is one where an external event causes the exception; this can effectively occur at any time and is not dependent on the instruction flow. A precise exception is where the cause is precisely defined and is usually recoverable. A memory page fault is a good example of this. An imprecise exception is usually a catastrophic failure, where the processor cannot continue processing or allow a particular program or task to continue. A system reset or memory fault while accessing the vector table falls into this category.

Synchronous precise

All instruction–caused exceptions are handled as synchronous precise exceptions. When such an exception is encountered during program execution, the address of either the faulting instruction or the one after it is stored in SRR0. The processor will have completed all the preceding instructions, however, this does not guarantee that all memory accesses caused by these instructions are complete. The faulting instruction will be in an indeterminate state, i.e. it may have started and partially or completely completed. It is up to the exception handler to determine the instruction type and its completion status using the information bits in the DSISR and FPSCR registers.

Synchronous imprecise

This is generally not supported within the PowerPC architecture and is not present on the MPC601, MPC603 or MCP604 implementations. However, the PowerPC architecture does specify the use of synchronous imprecise handling for certain floating point exceptions and so this category may be implemented in future processor designs.

Asynchronous precise

This exception type is used to handle external interrupts and decrementer–caused exceptions. Both can occur at any time within the instruction processing flow. All instructions being processed before the exception are completed, although there is no guarantee that all the memory accesses have completed. SRR0 stores the address of the instruction that would have been executed if no interrupt had occurred.

These exceptions can be masked by clearing the EE bit to zero in the MSR. This forces the exceptions to be latched but not acted on. This bit is automatically cleared to prevent this type of interrupt causing an exception while other exceptions are being processed.

The number of events that can be latched while the EE bit is zero is not stated. This potentially means that interrupts or decrementer exceptions could be missed. If the latch is already full, any subsequent events are ignored. It is therefore recommended that the exception handler performs some form of handshaking to ensure that all interrupts are recognised.

Asynchronous imprecise

Only two types of exception are associated with this: system resets and machine checks. With a system reset all current processing is stopped, all internal registers and mem-ories are reset; the processor executes the reset vector code and effectively restarts processing.

The machine check exception is only taken if the ME bit of the MSR is set. If it is cleared, the processor enters the checkstop state.

Recognising an exception

Recognising an exception in a superscalar processor, especially one where the instructions are executed out of program order, can be a little tricky – to say the least. The PowerPC architecture handles synchronous exceptions (i.e. those caused by the instruction stream) in strict program order, even though instructions further on in the program flow may have already generated an exception. In such cases, the first exception is handled as if the following instructions have never been executed and the preceding ones have all completed.

There are occasions when several exceptions can occur at the same time. Here, the exceptions are handled on a priority basis using the priority scheme shown in the table overleaf. There is additional priority for synchronous precise exceptions because it is possible for an instruction to generate more than one exception. In these cases, the exceptions would be handled in their own priority order as shown below.

Class	Priority	Description
Async imprecise	1	System reset
	2	Machine check
Sync precise	3	Instruction dependent
Async precise	4	External interrupt
	5	Decrementer interrupt

Exception class priority

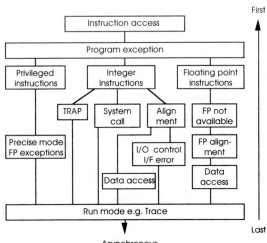

Precise exceptions priority

If, for example, with the single-step trace mode enabled, an integer instruction executed and encountered an alignment error, this exception would be handled before the trace exception. These synchronous precise priorities all have a higher priority than the level 4 and 5 asynchronous precise exceptions, i.e. the external interrupt and decrementer exceptions.

When an exception is recognised, the continuation instruction address is stored in SRR0 and the MSR is stored in SRR1. This saves the machine context and provides the interrupted program with the ability to continue. The continuation instruction may not have started, or be partially or fully complete, depending on the nature of the exception. The FPSCR and DSISR registers contain further diagnostic information for the handler.

When in this state, external interrupts and decrementer exceptions are disabled. The EE bit is cleared automatically to prevent such asynchronous events from unexpectedly causing an exception while the handler is coping with the current one.

It is important to note that the machine status or context which is necessary to allow normal execution to continue is automatically stored in SRR0 and SRR1 – which overwrites the previous contents. As a result, if another exception occurs during an exception handler execution, the result can be catastrophic: the exception handler's machine status information in SRR0 and SRR1 would be overwritten and lost. In addition, the status information in FPSCR and DSISR is also overwritten. Without this information, the handler cannot return to the original program. The new exception handler takes control, processes its exception and, when the rfi instruction is executed, control would is passed back to the first exception handler. At this point, this handler does not have its own machine context information to enable it to return control to the original program. As a result the system will, at best, have lost track of that program; at worst, it will probably crash.

This is not the case with the stack-based exception handlers used on CISC processors. With these architectures, the machine status is stored on the stack and, providing there is sufficient stack available, exceptions can safely be nested, with each exception context safely and automatically stored on the stack.

It is for this reason that the EE bit is automatically cleared to disable the external and decrementer interrupts. Their asynchronous nature means that they could occur at any time and if this happened at the beginning of an exception routine, that routine's ability to return control to the original program would be lost. However, this does impose several constraints when programming exception handlers. For the maximum performance in the exception handler, it cannot waste time by saving the machine status information on a stack or elsewhere. In this case, exception handlers should prevent any further exceptions by ensuring that they:

- reside in memory and not be swapped out.

- have adequate stack and memory resources and not cause page faults.

- do not enable external or decrementer interrupts.

- do not cause any memory bus errors.

For exception handlers that require maximum performance but also need the best security and reliability, they should immediately save the machine context, i.e. SRR registers FPSCR and DSISR, preferably on a stack before continuing execution.

In both cases, if the handler has to use or modify any of the user programming model, the registers must be saved prior to modification and they must be restored prior to passing control back. To minimise this process, the supervisor model has access to four general purpose registers which it can use independently of the general purpose register file in the user programming model.

Enabling exceptions

Some exceptions can be enabled and disabled by the supervisor by programming bits in the MSR. The EE bit controls external interrupts and decrementer exceptions. The FE0 and FE1 bits control which floating point exceptions are taken. Machine check exceptions are controlled via the ME bit.

Returning from exceptions

As mentioned previously, the rfi instruction is used to return from the exception handler to the original program. This instruction synchronises the processor, restores the instruction address and machine state register and the program restarts.

The term 'restart' is important and has some implications. Unlike many CISC processors (for example, the MC68000, MC68020 and MC68030) the instruction does not continue; it is restarted from the beginning. If the exception ocurred in the middle of the instruction, the restart repeats the initial action. For many instructions this may not be a problem – but it can lead to some interesting situations concerning memory and I/O accesses.

If the instruction is accessing multiple memory locations and fails after the second access, the first access will be repeated. The store multiple word is a good example of this, where the contents of several registers are written out to memory. If the target address is an I/O peripheral, an unexpected repeat access may confuse it.

The vector table

Once an exception has been recognised, the program flow changes to the associated exception handler contained in the vector table.

The vector table is a 16 kbyte block (0 to $3FFF) that is split into 256 byte divisions. Each division is allocated to a particular exception or group of exceptions and contains the exception handler routine associated with that exception. Unlike many other architectures, the vector table does not contain pointers to the routines but the actual instruction sequence themselves. If the handler is too large to fit in the division, a branch must be used to jump to its continuation elsewhere in memory.

The table can be relocated by changing the EP bit in the machine state register (MSR). If cleared, the table is located at $0000000. If the bit is set to one (its state after reset) the vector table is relocated to $FFF00000. Obviously, changing this bit before moving the vector table can cause immense problems!

Identifying the cause

Most programmers will experience exception processing when a program has crashed or a cryptic message is returned from a system call. The exception handler can provide a lot of information about what has gone wrong and the likely cause. In this section, each exception vector is described and an indication of the possible causes and remedies given.

The first level investigation is the selection of the appropriate exception handler from the vector table. However, the exception handler must investigate further to find out the exact cause before trying to survive the exception. This is done by checking the information in the FPSCR, DSISR, DAR and MSR registers, which contain different information for each particular vector.

The vector descriptions apply to both the MPC601 and the MPC603, unless otherwise stated.

Vector Offset (hex)	Exception	
0 0000	Reserved	
0 0100	System Reset	Power-on, Hard & Soft Resets
0 0200	Machine Check	Eabled through MSR [ME]
0 0300	Data Access	Data Page Fault/Memory Protection
0 0400	Instruction Access	Instr. Page Fault/Memory Protection
0 0500	External Interrupt	INT
0 0600	Alignment	Access crosses Segment or Page
0 0700	Program	Instr. Traps, Errors, Illegal, Privileged
0 0800	Floating-Point Unavailable	MSR[FP]=0 & F.P. Instruction encountered
0 0900	Decrementer	Decrementer Register passes through 0
0 0A00	Reserved	
0 0B00	Reserved	
0 0C00	System Call	'sc' Instruction
0 0D00	Trace	Single-step instruction trace
0 0E00	Floating -Point Assist	A floating point exception

The basic PowerPC vector table

'0 0000' reserved

The first vector in the table is normally associated with the first instructions the processor executes when it comes out of reset. This is not the case with the PowerPC, where this vector is reserved for future use and not used to supply the first instructions.

Supervisor-Level SPRs

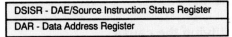

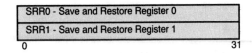

Supervisor-Level

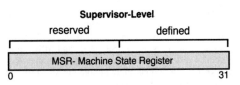

Registers involved in handling exceptions

'0 0100' system reset

This is where the processor starts to execute its first instructions. All data translation is disabled and the addresses it uses should be available in physical memory. This handler is a little strange in that it often performs the processor configuration and downloading of the operating system, which is strictly not speaking the response to an exception and is usually only performed once – i.e. when the processor starts up. When the processor comes out of a hard reset, the EP bit in the MSR is set to 1 and therefore the vector table must be located at $FFF0000. Needless to say, all the registers, TLBs and caches are also reset and any previous data is lost.

The MPC601 and the MPC603 both support a soft reset. This is similar to a hard reset, in that the processing is broken and the processor may have to be restarted, but the MSR EP bit is not reset and so the system reset vector is taken from the current location for the vector table. In addition, it may be possible to restart the processor either from the beginning or from the next instruction it was going to execute before the soft reset occurred. The problem with the second option is that there is no guarantee the renewed processing stream will continue as if nothing had happened.

The real use of the software reset is in providing some level of non-maskable interrupt where the current processing can be interrupted and the machine state investigated. However, unlike its CISC equivalents, it is an imprecise exception and therefore recovery cannot be guaranteed. With an MC68020, a level 7 non-maskable interrupt can be generated to put the processor into a debugger and the current status can be investigated or single stepped. The go command in such debuggers then starts the processor executing as normal. With the imprecise exception that the soft reset generates, the processor can be made to start processing through an equivalent command, but there is no guarantee that it will do so. It may continue processing correctly, or with corrupt data, or just crash. There is no way of predicting which route it will take.

If a system has hung up, a soft reset is the only way of getting diagnostic information from the system. As such, every system should have a diagnostic routine embedded into this exception handler to handle the soft reset case and provide this facility.

'0 0200' machine check

The machine check exception is implementation dependent in that the level of checking performed is processor dependent. This exception is normally caused by an external memory fault such as TEA* on the MPC601 and the MPC603 or by asserting the MCP pin.. The processor performs its machine checking and, providing the exception is enabled, goes to that vector. If the exception is disabled, the processor goes into its checkstop condition, where the external bus goes into a high impedance state and effectively stops. Once in this state, a hard reset is the only way out.

The ability to take an exception allows an operating system to attempt a recovery although, in most cases, this is limited as the exception is imprecise. If the condition that caused the exception does not prevent continuation, instruction execution will continue using the machine check vector. With the MPC603, the RI bit of the MSR is also set to indicate that the exception is non-recover-able. All that can be done in these cases is to log the event and try to continue. If the system was running multiple tasks or programs, the offending program could be removed, the task scheduler restarted and control passed to a different task.

The MPC601 uses the HID0 register to enable and identify the cause of this exception.

Bit	Name	Description
0	CE	Master checkstop enable.
1	S	Microcode checkstop detected if set.
2	M	Double machine check detected if set.
3	TD	Multiple TLB hit checkstop if set.
4	CD	Multiple cache hit checkstop if set.
5	SH	Sequencer time out checkstop if set.
6	DT	Dispatch time out checkstop if set.
7	SA	Bus address parity error if set.
8	BD	Bus data parity error if set.
9	CP	Cache parity error if set.
10	IU	Invalid microcode instruction if set.
11	PP	I/O controller interface access protocol error if set.

The MPC601 HID0 register checkstop conditions

'0 0300' data access

This handler is invoked if a data access cannot be performed. The effective address of the offending access is placed in the DAR and the DSISR describes the problem.

Bit	Description
0 MPC603	I/O controller interface exception. The MPC601 reserves this bit and uses the ' 0 0A00' vector instead.
1	Translation not found in the BAT or the primary or secondary hashed PTEGs.
2-3	Cleared.
4	Memory protection fault.
5 MPC601	lwarx, stwcx or lscbx access to I/O controller interface space.
5 MPC603	Set if the access is to an I/O segment by an eciwx, ecowx, lwarx or stwcx instruction. It is also set if the address is marked as write through and is accessed by an eciox or ecowx instruction.
6	Set for a store, cleared for a load.
7-8	Cleared.
9 MPC601	DABR breakpoint
9 MPC603	Cleared.
10	No segment translation for effective address.
11	Attempt to execute eciwx or ecowx instructions with the E bit of the EAR register cleared.
12-31	Cleared.

Data access exception DSISR codings

There are several causes:

•   The effective address has caused a page fault.

In this case, the handler must go out and create the page table entry and, if necessary, swap some pages of memory out to disk to free up enough spare pages to support the access. It must also ensure that any TLB entries are invalidated. If the processor supports hardware tablewalking, the handler simply needs to execute a rfi to restart the access. The processor then searches its BATs and TLBs for the PTE and not find it, as before, but this time successfully fetch it from memory.

With a software table walk, as used with the MPC603, the handler may have to load the PTE entry into the data TLB or set up the DBAT to handle it before allowing the access to restart.

•   The type of memory addressed is not supported by the instruction.

This is a programming error and therefore little can be done except remove the task and return an error message explaining the problem. The access violates the memory protection attributes set by the PP bits and either KS or KU.

This is often caused when a coding problem has created a wrong address. It is often the first sign that something has gone wrong with a task. Although the cause may have been a lot earlier in the instruction sequence, the resulting instructions have been legal although incorrect. At some point, a register or address becomes corrupted and changes the instruction from incorrect to illegal and causes an access exception. If the address is nothing to do with the task but the instruction making the access is, this is likely to a symptom rather than a cause. If the instruction and address are in the operating system or somewhere else, a wrong parameter is often the possible cause.

Other causes include the wrong access permissions for shareable segments. In this case, the segment is shared between several tasks but the incorrect access permissions have been given to one or more of the tasks, so a task may only have read permission when it assumes that it has read and write access. When it writes to the location, it gets a memory protection error. If the address is associated with a task, check these permissions.

- The execution of eciwx or ecowx instructions while the E bit in the EAR register is cleared, thus disallowing its use.

This is either a program error or corrupt instructions where the corruption has turned the original instructions into one of these two.

- The effective address matches that stored in the data address breakpoint register and therefore has generated a breakpoint, resulting in this exception. This is often used as a debugging tool (MPC601 only, not supported on the MPC603).

- Incorrect addressing with the eciwx type of instructions.

Bit 5 has two slightly different interpretations, depending whether the processor is an MPC601 or an MPC603. With the MPC601, this indicates the use of lwarx, stwcx and lscbx instructions with non-memory forced I/O controller interface segment. With the MPC603, the list of instructions that can cause this problem is different and the address faults can include having a write-through attribute associated with the address in an ecowx or eciwx instruction. These faults are not common.

'0 0400' instruction access

Bit	Description
0	Cleared
1	Translation not found in the BAT or the primary or secondary hashed PTEGs.
2	Cleared.
3	Set if the access was to an I/O controller interface segment.
4	Memory protection fault.
5-9	Cleared
10	No segment translation for effective address.
11-15	Cleared.
16-31	Loaded from the MSR.

Instruction access exception SSR1 codings

The causes of this exception are similar to data access except that the fault occurred in fetching an instruction.

- The effective address has caused a page fault.

In this case, the handler must go out and create the page table entry and, if necessary, swap some pages of memory out to disk to free up spare pages to support the access. It must also ensure that any TLB entries are invalidated. If the processor supports hardware tablewalking, the handler simply needs to execute a rfi to restart the access. The processor then searches its BATs and TLBs for the PTE. It may not find it, but this time successfully fetch it from memory.

With a software table walk, the handler must load the PTE entry into the data TLB or set up the DBAT to handle it before allowing the access to restart.

• The type of memory addressed is not supported by the instruction.

This is a programming error and therefore little can be done except remove the task and return an error message explaining the problem.

The access violates the memory protection attributes set by the PP bits and KS, KU or the address was located in the I/O controller interface segment.

This is often caused by a coding problem creating a wrong address. It is often the first sign that something has gone wrong with a task. Although the cause may have been a lot earlier in the instruction sequence, the resulting instructions have been legal although incorrect. At some point, a register or address becomes corrupted and changes the instruction from being incorrect to illegal and causes an access exception. If the address is nothing to do with the task but the instruction making the access is, this is likely to be a symptom rather than a cause. If the instruction or address is in the operating system or somewhere else, a wrong parameter is the possible cause.

Other causes include the wrong access permissions for shareable segments. In this case, the segment is shared between several tasks but the incorrect access permission have been given to one or more of the tasks. That task may only have read permission when it assumes that it has read and write access. When it writes to the location, it gets a memory protection error If the address is associated with a task, then check these permissions.

'0 0500' external interrupt

This exception is caused by the generation of an external interrupt asserting the INT* signal. The handler normally investigates the external hardware, identifies the cause, services the request and clears the interrupt. This should remove the signal and prevent false triggering.

When the exception handler is started, external and decrementer interrupts are disabled. Care must be taken not to re-enable them without having saved the machine context and cleared the interrupt source. This problem can be a difficult to recognise, although repeated entry into the handler with phantom interrupts can be a common system fault. If the external interrupt is enabled (the EE bit in the MSR) and the source has not been

cleared, the interrupt may be seen again and the exception re-
entered where the scenario is repeated, with the previous machine
context continually being lost. The result is a processor that loops
around the interrupt exception until some other fault diverts it to
a higher priority exception. In the end, the system has effectively
crashed.

A variation on this scenario is where the interrupt is enabled
after the software source has been cleared but the phantom
interrupt is still seen. The reason for this is hardware dependent
and is caused by the difference in time taken to clear the external
source and change the EE bit internally. Due to these time
differences, the external write to the hardware is not completed
before the EE bit is changed and, thus, the interrupt is still seen and
the exception routine incorrectly restarted.

It is recommended that the hardware source is cleared and the
processor synchronised before changing the EE bit within a
handler. This ensures that the memory access has at least completed before enabling interrupts again. However, the story does
not end there: depending on how the hardware is designed,
additional delays may be required to ensure successful completion. Although the processor may have completed the memory
access needed to clear the interrupt source, the peripheral may
require additional time to propagate the change through its hardware. A software loop that polls until the status change has been
made in the peripheral is often the most reliable way of making
sure that the source is removed.

The rfi instruction does synchronise the processor before
restoring the machine context (and enabling external interrupts)
but additional time may still be needed to compensate for the
propagation delay.

'0 0600' alignment

Alignment exceptions are generally caused by accesses that
go across page, segment, half word, word or double word bounda-
ries. They can also be caused when a floating point operand is in
an I/O controller interface segment. The typical causes are shown
in the table.

Bit	Description
0 *MPC603*	I/O controller interface exception.
0 *MPC601*	The MPC601 reserves this bit and uses the ' 0 0A00' vector instead.
1	Translation not found in the BAT or the primary or secondary hashed PTEGs.
2-3	Cleared.
4	Memory protection fault.
5	lwarx, stwcx or lscbx access to I/O controller interface space.
6	Set for a store, cleared for a load.
7-8	Cleared.
9	DABR breakpoint
10	No segment translation for effective address.
11	Attempt to execute eciwx or ecowx instruc- tions with the E bit of the EAR register cleared.
12-31	Cleared.

Data access exception DSISR codings

Access	Alignment fault cause.
Byte	Does not cause an alignment fault at all.
Half word	Only when the address ends in $FFF.
Word	Only when the address ends in a value from $FFD to $FFF.
Double word	Only when the address ends in a value from $FF9 to $FFF.
lscbx instruction	Causes an exception when crossing page boundary.
String instructions	Can cause an exception crossing page and 256 Mbyte segment boundary.
Operands	Causes an exception when crossing a page boundary.
Instruction	Causes an exception if any part crosses a 256 Mbyte boundary and address translation is disabled.
FP loads & stores	Causes an exception if the address is in the I/O controller interface segment or if address translation is dis-abled and the address crosses a 256 Mbyte boundary.
Little-endian mode	Any execution of string instructions or data that is not correctly aligned.
I/O controller I/F	Controller error can cause an alignment exception (MPC603 only).

Data access exception DSISR codings

The MPC601 handles I/O controller errors differently: it uses the '0 0A00' vector instead and reserves bit 0.

'0 0700' program

A program exception is caused by one of four conditions:

- A floating point exception

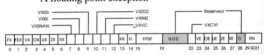

Floating point status and control register

This requires that the exception has been enabled in the first place by setting bits in the FPSCR register. This register is also used to identify the cause of the floating point exception. This handler does not handle exceptions caused by the floating point being unavailable – these are handled by the next exception handler ' 0 0800'.

Flag	Description
FX	Floating point exception flag. Set by every floating point instruction that causes an exception. This is a sticky bit and must be cleared by software.
FEX	Floating point enabled exception flag. Set by every floating point instruction that causes an enabled exception condition. If this is clear and FX is set, then the exception is one that the user has no control over. If both FX and FEX are set, then the exception has been caused by an exception that the user has explicitly enabled.

Flag	Description
VX	Floating point invalid operation exception flag. This is set if the exception cause was an invalid operation.
OX	Floating point overflow exception.
UX	Floating point underflow exception.
ZX	Floating point zero divide exception.
X	Floating point inexact exception.
VXSNAN	Invalid operation involving a SNaN. This is a sticky bit and must be cleared by software.
VXISI	Invalid operation involving infinity minus infinity. This is a sticky bit and must be cleared by software.
VXIDI	Invalid operation involving infinity divided by infinity. This is a sticky bit and must be cleared by software.
VXZDZ	Invalid operation involving zero divided by zero. This is a sticky bit and must be cleared by software.
VXIMZ	Invalid operation involving infinity divided by zero. This is a sticky bit and must be cleared by software.
VXVC	Invalid operation involving an invalid compare. This is a sticky bit and must be cleared by software.
FR	The floating point fraction has been rounded.
I	The floating point fraction is inexact.
FPRF	Floating point result flags.
VXSOFT	Floating point invalid operation. This flag allows software to set a invalid operation condition that may not be associated with floating point operations such as the square root of a negative number. Not implemented in the MPC601.
VXSQRT	Invalid operation involving square root. This is a sticky bit and must be cleared by software. This is not implemented in the MPC601.
VXCVI	Invalid operation involving integer conversion. This is a sticky bit and must be cleared by software.
VE	Enable invalid operation exceptions.
OE	Enable overflow exceptions.
UE	Enable underflow exceptions.
ZE	Enable zero divide exceptions.
XE	Enable inexact exceptions.
RN	Floating point rounding control 00 round to nearest 01 round towards zero 10 round towards +infinity 11 round towards -infinity

Floating point status and control register flags

Result flags	Description
(bits 15-19)	
10001	Quiet NaN
01001	- infinity
01000	- normalised number
11000	- denormalised number
10010	- zero
00010	+ zero
10100	+ denormalised number
00100	+ normalised number
00101	+ infinity

FPSCR FPRF result flag settings

- Execution of an illegal instruction.

An illegal instruction either does not exist within the architecture or is not supported by the processor. This includes the MPC601 POWER architecture instructions, which are included in its instruction set but not supported in the PowerPC architecture or other members of the processor family. Optional instructions which act as no-ops do not cause this exception.

The handler has several choices in handling this problem. If the instruction does not exist, there is little that can be done as this is a programming error. Be careful in assuming that this is the root cause of the problem — often it is the symptom and the cause occurred in an instruction before it. If the instruction is legal but not implemented, the handler can simulate it if required. This is the method used to provide POWER instruction support for those instructions that are not supported in the PowerPC architecture and/or processor implementation. The instruction is detected as illegal and passed to this handler which then simulates it. There are some limitations concerning this support: simulating 64 bit address and data PowerPC architecture could be done, but is difficult and costly. One of the main problems is that the registers are not implemented in the MPC601 and MPC603 as a full 64 bit set and so the additional information would need to be simulated and checked after almost every instruction execution.

Illegal instructions are sometimes used to provide breakpoints: an illegal instruction is substituted at the breakpoint address. When it is executed and the exception taken, the handler switches into a debugger mode and displays the breakpoint status

- Privileged instruction

This is caused by the user trying to access supervisor resources while the PR bit of the MSR is set to one. Again, the exception can be handled as a true error or the handler could simulate the request for the user, although this can be dangerous and requires careful consideration.

This problem can also be caused from within a supervisor program if the PR bit is set to 1 in error. A normal supervisor access would then be seen as an error.

- Trap instruction

In this case, the handler is called by successful matching of a condition within a trap instruction. This is often used as a method of making system calls.

'0 0800' floating point unavailable

If the FP bit of the MSR is cleared to zero and a floating point instruction executed, the attempt is aborted and this exception handler is invoked. This is often caused by a programming error. The address of the offending instruction is placed in SRR0 and it is therefore possible for the handler to simulate the floating point operation using software, if needed. In this way, it is possible to implement a different floating point environment in software and running at a far slower processing rate, yet using the same instructions and register set. The floating point would need to be enabled within the handler to allow the software simulation to access the floating point register file, and disabled when control is returned.

'0 0900' decrementer

This is similar to the external interrupt, in that it is controlled by the EE bit in the MSR. The exception is caused when the DEC register transitions from zero to one, either as a result of a clock or by software.

'0 0C00' system call

This is similar to the trap instruction with the program exception, in that it is called when the SC instruction is executed. It provides a communication channel between the user and the supervisor and is often used to request a system call.

'0 0D00' trace

This is not supported on the MPC601, which uses its run mode vector '0 2000' to provide similar support. Within the PowerPC architecture and the MPC603, this is used to process single stepping. If the SE bit of the MSR is set to one, this exception handler is called after every instruction completion. If the BE bit of the MSR is set to one and the instruction is a branch, this handler is also called.

In both cases, the handler has an opportunity to provide debug and status information. This facility is essential when single stepping or tracing through a program from within a debugger. These facilities would not normally be used in normal situations because of the performance impact. To provide the single stepping, each instruction is executed serially and must complete before the next one is dispatched. The superscalar execution is effectively disabled. As a result, throughput is greatly reduced.

'0 0E00' floating point assist

This is not supported on either the MPC601, MPC603 or the MPC604 but is defined in the PowerPC architecture. It is used to provide software assistance for floating point calculations.

'0 0A00' MPC601 I/O controller error

This vector is used on the MPC601 instead of the data access vector to handle I/O controller interface errors.

'0 0F00' MPC604 performance monitor

This exception is specific to the MPC604 and is used to obtain performance information. It is generated when the value in one of the performance monitor counter registers (PMC1 or PMC2) goes negative. The conditions that can cause this exception can be enabled or disabled by through bits in the monitor mode control register 0 (MMCR0).

The handler can extract information which it can then process to create a series of statistics about the system's performance.

'0 1000' MPC603 instruction translation miss

'0 1100' MPC603 data load translation miss

'0 1200' MPC603 data store translation miss

These three vectors are part of the support for the software table walking facility that memory management uses on the MPC603 to fetch page table entries from external memory. These exception handlers go out to memory using the primary and secondary hashed address, search for the page table entry, update the appropriate TLB and restart the access,

The three different vectors are used to provide and discriminate between instruction accesses, and data stores and loads.

'0 1300' MPC603/4 instruction breakpoint

This exception handler is called if the address in the IABR matches a dispatched instruction. It is provided to allow breakpoints to be easily implemented without having to use the illegal instruction method. The handler can jump into a debugger or pass the processor status to the user for debugging purposes.

'0 1400' MPC603/4 system management

This is a special form of interrupt which is caused when the system management interrupt pin SMI* is asserted. It is maskable by clearing the EE bit of the MSR register and allows a system manager to have a fast direct communication channel into the processor. This can be used to start system housekeeping jobs, synchronise the processor or timers, or inform it of some other event.

'0 2000' MPC601 run mode

This is specific to the MPC601 and covers some of the functions that the PowerPC architecture and MPC603 dedicate separate vectors to. It is called under the following conditions controlled by the HID1 register.

As can be seen from the HID1 register description, this vector is only called when the RM field is set to 10. In addition, the M field also controls the run mode and the level to which the address comparison is performed. With the full address comparison, processor performance is impaired. If only instruction breakpoints are needed, the run mode can be set to the limited instruction address compare which has less performance impact.

0	M	0000	RM	0000000	MISC
0 1	3 4	7 8 9	10	16 17	31

Reserved

The HID1 register

Field	Code	Description
M	000	Normal run mode
	001	Undefined — do not use!
	010	Limited instruction address compare.
	011	Undefined — do not use!
	100	Single instruction step
	101	Undefined — do not use!
	110	Full instruction address compare.
	111	Full data address compare.
RM	00	Hard stop.
	01	Soft stop - waits for system activity to stop.
	10	Trap to run mode exception vector '0 2000'.
	11	Reserved — do not use!
MISC	—	Bit 17 when high disables the broadcast of the tlbie instruction. All other bits in this field are reserved and should not be used.

The HID1 register bit descriptions

Other vectors

All other vectors that have not been explicitly mentioned in this chapter up to '0 3F00', are reserved and should not be used. It is good practice to have simple handlers installed at these locations in case of a severe system malfunction! A simple handler can at least register the error whereas no handler at all will give no information at all.

Chapter 6

The PowerPC instruction set

This chapter details the PowerPC instruction set, including the addressing modes, all the 32 bit and 64 bit derivatives, alternative mnemonics and the POWER architecture instructions supported by the MPC601. The instructions are ordered alphabetically.

The details cover the following topics:

- Opcode name

    This is the name of the opcode and reflects the action that the opcode performs. It is this name that has been used to alphabetically order the entries within this chapter.

- Syntax

    This is the syntax used by the opcode and typically consists of three entries although this is not always the case. The operands are explained in the decription.

- Derivatives

    This information indictes whether the opcode supports the three suffices '.', 'o', and 'o.'. The supported forms of the opcode are shown here and can be substituted within the syntax example to create all the supported forms. The suffices have the following actions:

    .   Update the condition register.

    o   Update the overflow bits.

    o.   Update both the overflow bits and condition register.

- Processor support

    This entry defines which architecture and/or processors supports this instruction.

- Description

    This gives a detailed explanation and other useful information about the instruction.

Power PC addressing modes

There are four fundamental addressing modes supported in the PowerPC architecture.

Register index with no update

With this mode, two registers are added together to create the address that is used by the instruction. This address is logical and is translated, if needed, by the memory management unit.

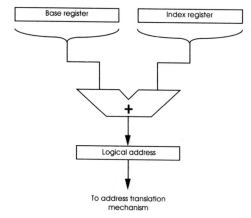

Register index with no update

Register index addressing with update

This is the same as the previous example, except that base register is updated with the effective address. This is an extremely useful addition as it allows a pointer to be moved through a data structure without having to continually re-calculate and store its new position.

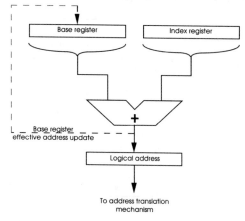

Register index addressing with update

Immediate with no update

This addressing mode uses a signed 16 bit immediate value, instead of the index register, to calculate the new effective address.

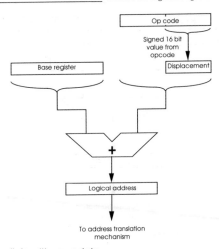

Immediate with no update

Immediate addressing with update

This is similar to the previous addressing mode, except that the base register is updated with the effective address. This mode is very good for controlling stacks. The base register is used as the stack pointer and the signed immediate values used to pop and push data onto the stack as needed.

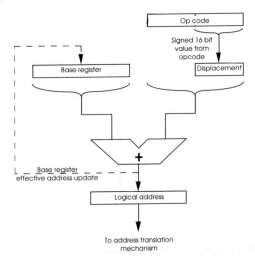

Immediate addressing with update

Absolute
abs   rD,rA
abs abs. abso abso.
• POWER • MPC601.

The absolute value |(rA)|is placed into register D. If register A contains the most negative number (i.e., $80000000), the result of the instruction is the most negative number and sets the XER[OV] bit only if the overflow is enabled.

Add
add   rD,rA,rB
add add. addo addo.
• 32 bit PowerPC

The contents of registers A and B are added together and the result is placed into register D. Carrying is ignored.

The normal format of this calculation is A+B=C. To simulate a diadic instruction set which will use the A=B=B format, assign one of the source registers as the destination.

Add carrying
addc   rD,rA,rB
addc addc. addco addco.
• 32 bit PowerPC

The contents of registers A and B are added together and the result is placed into register D. In this case, carrying is detected and the XER[CA] bit will be set if a carry occurs as a result of this calculation.

The normal format of this calculation is A+B=C. To simulate a diadic instruction set which will use the A=B=B format, assign one of the source registers as the destination.

Add Extended
adde   rD,rA,rB
adde adde. addeo addeo.
• 32 bit PowerPC

The contents of registers A and B are added together with the carry bit XER[CA], and the result is placed into register D. In this case, carrying is detected and the XER[CA] bit will be set if a carry occurs as a result of this calculation.

The normal format of this calculation is A+B=C. To simulate a diadic instruction set which will use the A=B=B format, assign one of the source registers as the destination.

Add Immediate
addi. rD,rA,SIMM

addi.
• **32 bit PowerPC**

The contents of register A or zero is added to SIMM and the result is placed into register rD. If r0 is specified as rA, the value taken is not its contents but zero. The SIMM value is sign extended to make it a 32 bit integer prior to the addition.

This form can be used to load a register with a signed immediate value.

Add Immediate Carrying
addic rD,rA,SIMM

addic
• **32 bit PowerPC**

The contents of register A or zero is added to SIMM and the result is placed into register rD. If r0 is specified as rA, the value taken is not its contents but zero. If carrying occurs, the carry bit XER[CA] is updated. The SIMM value is sign extended to make it a 32 bit integer prior to the addition.

Add Immediate Carrying and Record
addic. rD,rA,SIMM

addic.
• **32 bit PowerPC**

The contents of register A or zero is added to SIMM and the result is placed into register rD. If r0 is specified as rA, the value taken is not its contents but zero. If carrying occurs, the carry bit XER[CA] is updated. In addition the condition register is updated. The SIMM value is sign extended to make it a 32 bit integer prior to the addition.

Although this instruction appears to be the same as the addic instruction with the . suffix to allow the update of the condition register, it has a different primary opcode and has therefore been treated as a separate case. The SIMM value prevents bit 31 from being used to encode the . suffix.

Add Immediate Shifted
addis. rD,rA,S!MM

addis.
• **32 bit PowerPC**

The contents of register A and the number (SIMM || $0000) are added and the result is placed into register D.

This instruction is used to load the upper 16 bits of a register with an immediate value. This instruction, in conjunction with ori, can be used to load an immediate 32 bit value into a register. The operation is done in two stages:

    addis. rX,r0,SIMM1
    ori   rD,rX,SIMM2

where SIMM1 is the number represented by the top 16 bits and SIMM2 is the number represented by the lower sixteen bits. The result from the first add must be or'ed into the second calculation otherwise the upper sixteen bits will be lost.

Add to Minus One
addme rD,rA

addme addme. addmeo addmeo.
• **32 bit PowerPC**

The contents of register A, the carry bit XER(CA) and $FFFFFFFF are added and the result is placed into register rD.

Add to Zero Extended
addze rD,rA

addze addze. addzeo addzeo.
• **32 bit PowerPC**

The contents of register A and the carry bit XER(CA) are added and then placed into register rD.

AND
and rA,rS,rB

and
• **32 bit PowerPC**

The contents of register S is ANDed with the contents of register B and the result is placed into register A.

AND with complement
andc   rA,rS,rB

andc andc.
• **32 bit PowerPC**

___

The contents of register S is ANDed with the one's complement of the contents of register B and the result is placed into register A.

AND Immediate
andi   rA,rS,UIMM

andi andi.
• **32 bit PowerPC**

___

The contents of register S is ANDed with $0000 || UIMM and the result is placed into register A.

Remember that the UIMM is only 16 bits long and that the upper bits within the register are effectively cleared by ANDing them with $0000. To AND the upper bits and clear the lower ones, use the andis instruction.

AND Immediate Shifted
andis   rA,rS,UIMM

andis andis.
• **32 bit PowerPC**

___

The contents of register S is ANDed with UIMM ||$0000 and the result is placed into register A.

This is the opposite of the andi instruction where the UIMM is ANDed with the upper 16 bits of the register and the lower 16 bits are effectively cleared.

Branch
b   imm addr

b
• **32 bit PowerPC**

___

Branch to the address computed as the sum of the immediate address and the address of the current instruction. The branch is always taken.

This is the equivalent to the jump type of instruction found in CISC instruction sets. Although described as a branch, the return address is not stored. If this is needed, use the bl instruction instead.

Branch Absolute
ba imm addr
ba
• **32 bit PowerPC**

Branch to the absolute address specified. The branch is always taken.

This is the equivalent to the jump type of instruction found in CISC instruction sets. Although described as a branch, the return address is not stored. If this is needed, use the bla instruction instead.

Branch Conditional
bc BO,BI,target addr
bc
• **32 bit PowerPC**

Branch conditionally to the address computed as the sum of the immediate address and the address of the current instruction. The BI operand specifies the bit in the condition register (CR) to be used as the condition of the branch. The BO operand is used as described:

BO	y=0	y=1	Description
0000y	0	1	Decrement the count register, then branch if the new value is not equal to 0 and the condition is FALSE.
0001y	2	3	Decrement the count register, then branch if the new value is equal to 0 and the condition is FALSE.
001zy	4	5	Branch if the condition is FALSE.
0100y	8	9	Decrement the count register, then branch if the new value is not equal to 0 and the condition is TRUE.
0101y	10	11	Decrement the count register, then branch if the new value is equal to 0 and the condition is TRUE.
011zy	12	13	Branch if the condition is TRUE.
1z00y	16	17	Decrement the count register, then branch if the new value is not equal to 0.
1z01y	18	19	Decrement the count register, then branch if the new value is equal to 0.
1z1zz	20	20	Branch always.

where z = 0 to create a valid field.

Although described as a branch, the return address is not stored. If this is needed, use the blc instruction instead.

Branch Conditional Absolute
bca   BO,BI,target addr

bca
• **32 bit PowerPC**
___

Branch conditionally to the absolute address specified. The BI operand specifies the bit in the condition register (CR) to be used as the condition of the branch. The BO operand is used as described:

BO	y=0	y=1	Description
0000y	0	1	Decrement the count register, then branch if the new value is not equal to 0 and the condition is FALSE.
0001y	2	3	Decrement the count register, then branch if the new value is equal to 0 and the condition is FALSE.
001zy	4	5	Branch if the condition is FALSE.
0100y	8	9	Decrement the count register, then branch if the new value is not equal to 0 and the condition is TRUE.
0101y	10	11	Decrement the count register, then branch if the new value is equal to 0 and the condition is TRUE.
011zy	12	13	Branch if the condition is TRUE.
1z00y	16	17	Decrement the count register, then branch if the new value is not equal to 0.
1z01y	18	19	Decrement the count register, then branch if the new value is equal to 0.
1z1zz	20	20	Branch always.

where z = 0 to create a valid field.

Although described as a branch, the return address is not stored. If this is needed, use the bcla instruction instead.

Branch Conditional to Count Register
bcctr   BO,BI

bcctr
• **32 bit PowerPC**
___

Branch conditionally to the address specified in the count register. The BI operand specifies the bit in the condition register (CR) to be used as the condition of the branch. The BO operand is used as described:

BO	y=0	y=1	Description
0000y	0	1	Decrement the count register, then branch if the new value is not equal to 0 and the condition is FALSE.
0001y	2	3	Decrement the count register, then branch if the new value is equal to 0 and the condition is FALSE.
001zy	4	5	Branch if the condition is FALSE.
0100y	8	9	Decrement the count register, then branch if the new value is not equal to 0 and the condition is TRUE.
0101y	10	11	Decrement the count register, then branch if the new value is equal to 0 and the condition is TRUE.
011zy	12	13	Branch if the condition is TRUE.
1z00y	16	17	Decrement the count register, then branch if the new value is not equal to 0.
1z01y	18	19	Decrement the count register, then branch if the new value is equal to 0.
1z1zz	20	20	Branch always.

where z = 0 to create a valid field.

Although described as a branch, the return address is not stored. If this is needed, use the bcctrl instruction instead.

The "decrement and test CTR" option – BO[2]=0 – should not be specified as it generates an invalid form of the instruction. With the MPC601, this variation is executed by using the decremented count register for the testing with zero, and using the non-decremented version of the count register as the target address. The best advice for complete compatibility, is not to use this varient at all.

Branch Conditional to Count Register then Link
bcctrl BO,BI

bcctrl

* **32 bit PowerPC**

Branch conditionally to the address specified in the count register. The instruction address following this instruction is placed into the link register. The BI operand specifies the bit in the condition register (CR) to be used as the condition of the branch. The BO operand is used as described:

BO	y=0	y=1	Description
0000y	0	1	Decrement the count register, then branch if the new value is not equal to 0 and the condition is FALSE.

BO	y=0	y=1	Description
0001y	2	3	Decrement the count register, then branch if the new value is equal to 0 and the condition is FALSE.
001zy	4	5	Branch if the condition is FALSE.
0100y	8	9	Decrement the count register, then branch if the new value is not equal to 0 and the condition is TRUE.
0101y	10	11	Decrement the count register, then branch if the new value is equal to 0 and the condition is TRUE.
011zy	12	13	Branch if the condition is TRUE.
1z00y	16	17	Decrement the count register, then branch if the new value is not equal to 0.
1z01y	18	19	Decrement the count register, then branch if the new value is equal to 0.
1z1zz	20	20	Branch always.

where z = 0 to create a valid field.

The "decrement and test CTR" option – BO[2]=0 – should not be specified as it generates an invalid form of the instruction. With the MPC601, this variation is executed by using the decremented count register for the testing with zero, and using the non-decremented version of the count register as the target address. The best advice for complete compatibility, is not to use this varient at all.

If the link register is subsequently modified without explicitly saving the previous value, the return address of one of the calls will be lost, with potentially disastrous results.

In other words, the link register does not act as a stack where the return addresses are neatly stacked and can be unstacked in order.

Branch Conditional then Link
bcl   BO,BI,target addr
bcl
• 32 bit PowerPC

Branch conditionally to the address computed as the sum of the immediate address and the address of the current instruction. The instruction address following this instruction is placed into the link register.The BI operand specifies the bit in the condition register, (CR) to be used as the condition of the branch. The BO operand is used as described:

BO	y=0	y=1	Description
0000y	0	1	Decrement the count register, then branch if the new value is not equal to 0 and the condition is FALSE.

BO	y=0	y=1	Description
0001y	2	3	Decrement the count register, then branch if the new value is equal to 0 and the condition is FALSE.
001zy	4	5	Branch if the condition is FALSE.
0100y	8	9	Decrement the count register, then branch if the new value is not equal to 0 and the condition is TRUE.
0101y	10	11	Decrement the count register, then branch if the new value is equal to 0 and the condition is TRUE.
011zy	12	13	Branch if the condition is TRUE.
1z00y	16	17	Decrement the count register, then branch if the new value is not equal to 0.
1z01y	18	19	Decrement the count register, then branch if the new value is equal to 0.
1z1zz	20	20	Branch always.

where z = 0 to create a valid field.

If the link register is subsequently modified without explicitly saving the previous value, the return address of one of the calls will be lost, with potentially disastrous results.

In other words, the link register does not act as a stack where the return addresses are neatly stacked and can be unstacked in order.

Branch Conditional Absolute then Link

bcla   BO,BI,target addr

bcla

▸ 32 bit PowerPC

Branch conditionally to the absolute address specified. The instruction address following this instruction is placed into the link register. The BI operand specifies the bit in the condition register (CR) to be used as the condition of the branch. The BO operand is used as described:

BO	y=0	y=1	Description
0000y	0	1	Decrement the count register, then branch if the new value is not equal to 0 and the condition is FALSE.
0001y	2	3	Decrement the count register, then branch if the new value is equal to 0 and the condition is FALSE.
001zy	4	5	Branch if the condition is FALSE.
0100y	8	9	Decrement the count register, then branch if the new value is not equal to 0 and the condition is TRUE.

BO	y=0	y=1	Description
0101y	10	11	Decrement the count register, then branch if the new value is equal to 0 and the condition is TRUE.
011zy	12	13	Branch if the condition is TRUE.
1z00y	16	17	Decrement the count register, then branch if the new value is not equal to 0.
1z01y	18	19	Decrement the count register, then branch if the new value is equal to 0.
1z1zz	20	20	Branch always.

where z = 0 to create a valid field.

If the link register is subsequently modified without explicitly saving the previous value, the return address of one of the calls will be lost, with potentially disastrous results.

In other words, the link register does not act as a stack where the return addresses are neatly stacked and can be unstacked in order.

Branch Conditional to Link Register

bclr   BO,BI

bclr

• **32 bit PowerPC**

Branch conditionally to the address in the link register. The BI operand specifies the bit in the condition register (CR) to be used as the condition of the branch.

The BO operand is used as described:

BO	y=0	y=1	Description
0000y	0	1	Decrement the count register, then branch if the new value is not equal to 0 and the condition is FALSE.
0001y	2	3	Decrement the count register, then branch if the new value is equal to 0 and the condition is FALSE.
001zy	4	5	Branch if the condition is FALSE.
0100y	8	9	Decrement the count register, then branch if the new value is not equal to 0 and the condition is TRUE.
0101y	10	11	Decrement the count register, then branch if the new value is equal to 0 and the condition is TRUE.
011zy	12	13	Branch if the condition is TRUE.
1z00y	16	17	Decrement the count register, then branch if the new value is not equal to 0.

BO	y=0	y=1	Description
1z01y	18	19	Decrement the count register, then branch if the new value is equal to 0.
1z1zz	20	20	Branch always.

where z = 0 to create a valid field.

The return address of the instruction after the bclr is not loaded into the link register.

This instruction can be used to conditionally return from a subroutine call where the return address is already in the link register. By specifying a condition that is always true, the instruction can be used in a similar way as a CISC processor would use rts to return from a subroutine.

Branch Conditional to Link Register then Link
bclrl   BO,BI

bclrl
• **32 bit PowerPC**

Branch conditionally to the address specified in the link register. The instruction address following this instruction is then placed into the link register. The BI operand specifies the bit in the condition register (CR) to be used as the condition of the branch. The BO operand is used as described:

BO	y=0	y=1	Description
0000y	0	1	Decrement the count register, then branch if the new value is not equal to 0 and the condition is FALSE.
0001y	2	3	Decrement the count register, then branch if the new value is equal to 0 and the condition is FALSE.
001zy	4	5	Branch if the condition is FALSE.
0100y	8	9	Decrement the count register, then branch if the new value is not equal to 0 and the condition is TRUE.
0101y	10	11	Decrement the count register, then branch if the new value is equal to 0 and the condition is TRUE.
011zy	12	13	Branch if the condition is TRUE.
1z00y	16	17	Decrement the count register, then branch if the new value is not equal to 0.
1z01y	18	19	Decrement the count register, then branch if the new value is equal to 0.
1z1zz	20	20	Branch always.

where z = 0 to create a valid field.

The return address of the instruction after the bclr is loaded into the link register.

This instruction can be used to conditionally go back to a subroutine call whose return address is already in the link register. Unlike the bclr instruction, the return address to return from this transfer is placed in the link register. By executing a bclr instruction, the flow would transfer back to the original stream. By specifying a condition that is always true, the instruction can be used in a similar way as a CISC processor would use to call a subroutine, but with the knowledge that the return address was stored in the link register. If the link register is subsequently modified without explicitly saving the previous value, the return address of one of the calls will be lost, with potentially disastrous results. In other words, the link register does not act as a stack where the return addresses are neatly stacked and can be unstacked in order.

Branch then Link
bl imm addr

bl
• **32 bit PowerPC**

___

Branch to the address computed as the sum of the immediate address and the address of the current instruction. This branch instruction saves the return address in the link register already for the return back to normal execution flow. If the link register is subsequently modified without explicitly saving the previous value, the return address of one of the calls will be lost, with potentially disastrous results.

In other words, the link register does not act as a stack where the return addresses are neatly stacked and can be unstacked in order.

Branch Absolute then Link
bla imm addr

bla
• **32 bit PowerPC**

___

Branch to the absolute address specified. This branch instruction saves the return address in the link register already for the return back to normal execution flow. If the link register is subsequently modified without explicitly saving the previous value, the return address of one of the calls will be lost, with potentially disastrous results.

In other words, the link register does not act as a stack where the return addresses are neatly stacked and can be unstacked in order.

Cache Line Compute Size
clcs   rD,rA

clcs
• POWER • MPC601.

This instruction places the cache line size specified by operand rA into register rD. The rA operand is encoded to specify which cache line size is required as follows:

    01100 Instruction cache line size (returns value of 64)
    01101 Data cache line size (returns value of 64)
    01110 Minimum line size (returns value of 64)
    01111 Maximum line size (return value of 64)
    All other encodings will return undefined values.
    For example:
    clcs r4,r15
    will put the value of 64 in register 4.

Clear Left and Shift Left Immediate
clrlslwi   rA,rS,b,n

clrlslwi clrlslwi.
• Alternative mnemomic

This alternative will clear the left most b bits of a register and then shift the result left by n bits. The value of b and n must be such that n b 31. It is equivalent of rlwinm rA,rS,n,b-n,31-n.

Clear Left Immediate
clrlwi   rA,rS,n

clrlwi clrlwi.
• Alternative mnemomic

This alternative will clear the left most n bits of a register. The value of n must be less than 32. It is equivalent of rlwinm rA,rS,0,n,31.

Clear Right Immediate
clrrwi   rA,rS,n

clrrwi clrrwi.
• Alternative mnemomic

This alternative will clear the right most n bits of a register. The value of n must be less than 32. It is equivalent of rlwinm rA,rS,0,0,31-n.

Compare
cmp   crfD,L,rA,rB

cmp
• 32 bit PowerPC

The contents of register A is compared with register B, treating the operands as signed integers. The result of the comparison is placed into the CR field specified by the operand crfD. This operand can have a value from 0 to 7.

If a series of compares are needed to check the upper and lower bounds of a numeric value, these can be performed using cmp instructions where the results are located in different CR fields. The results can either be examined individually or, logically combined using one of the condition register instructions before using the results as part of a conditional branch.

This instruction will support 64 bit operands as well as 32 bit values. If the L bit is set to 1, the instruction implies 64 bit operands. If this bit is zero, then 32 bit operands are used. The MPC601 ignores this bit and will always use 32 bit operands, but future versions of the PowerPC will support 64 bit operands and therefore, it is essential to correctly encode this bit for future compatibility. The L bit is bit 10 of the instruction.

Compare Immediate
cmpi   crfD,L,rA,SIMM

cmpi
• 32 bit PowerPC

The contents of register rA is compared with the 32 bit sign-extended value of the SIMM operand, treating the operands as 32 bit signed integers. The result of the comparison is placed into the CR field specified by operand crfD. This operand can have a value from 0 to 7.

If a series of compares are needed to check the upper and lower bounds of a numeric value, these can be performed using cmp instructions where the results are located in different CR fields. The results can either be examined individually or, logically combined using one of the condition register instructions before using the results as part of a conditional branch.

This instruction will support 64 bit operands as well as 32 bit values. If the L bit is set to 1, the instruction implies 64 bit operands. If this bit is zero, then 32 bit operands are used. The MPC601 ignores this bit and will always use 32 bit operands, but future versions of the PowerPC will support 64 bit operands and therefore, it is essential to correctly encode this bit for future compatibility. The L bit is bit 10 of the instruction.

Compare Logical
cmpl   crfD,L,rA,rB

cmpl
• 32 bit PowerPC

___

The contents of register rA is compared with register rB, treating the operands as unsigned integers. The result of the comparison is placed into the CR field specified by operand crfD. This operand can have a value from 0 to 7.

If a series of compares are needed to check the upper and lower bounds of a numeric value, these can be performed using cmp instructions where the results are located in different CR fields. The results can either be examined individually or, logically combined using one of the condition register instructions before using the results as part of a conditional branch.

This instruction will support 64 bit operands as well as 32 bit values. If the L bit is set to 1, the instruction implies 64 bit operands. If this bit is zero, then 32 bit operands are used. The MPC601 ignores this bit and will always use 32 bit operands, but future versions of the PowerPC will support 64 bit operands and therefore, it is essential to correctly encode this bit for future compatibility. The L bit is bit 10 of the instruction.

Compare Logical Immediate
cmpli   crfD,L,rA,UIMM

cmpli
• 32 bit PowerPC

___

The contents of register rA is compared with $0000 ‖ UIMM, treating the operands as 32 bit unsigned integers. The result of the comparison is placed into the CR field specified by operand crfD. This operand can have a value from 0 to 7.

If a series of compares are needed to check the upper and lower bounds of a numeric value, these can be performed using cmp instructions where the results are located in different CR fields. The results can either be examined individually or, logically combined using one of the condition register instructions before using the results as part of a conditional branch.

This instruction will support 64 bit operands as well as 32 bit values. If the L bit is set to 1, the instruction implies 64 bit operands. If this bit is zero, then 32 bit operands are used. The MPC601 ignores this bit and will always use 32 bit operands, but future versions of the PowerPC will support 64 bit operands and therefore it is essential to correctly encode this bit for future compatibility. The L bit is bit 10 of the instruction.

Compare Logical Word
cmplw   crfD,rA,rB

cmplw
- **Alternative mnemomic**

This alternative will perform a 32 bit unsigned compare rA with rB. If crfD is not specified, cr0 will be used by default. It is equivalent to cmpi crfD,0,rA,rB.

Compare Word
cmpw   crfD,rA,rB

cmpw
- **Alternative mnemomic**

This alternative will perform a 32 bit signed compare rA with rB. If crfD is not specified, cr0 will be used by default. It is equivalent to cmp crfD,0,rA,rB.

Compare Word Immediate
cmpwi   crfD,rA,SIMM

cmpwi
- **Alternative mnemomic**

This alternative will perform a 32 bit signed compare rA with an immediate value (SIMM). If crfD is not specified, cr0 will be used by default. It is equivalent to cmpi crfD,0,rA,SIMM.

Compare Word Logical Immediate
cmpwi   crfD,rA,UIMM

cmpwi
- **Alternative mnemomic**

This alternative will perform a 32 bit unsigned compare rA with an immediate value (UIMM). If crfD is not specified, cr0 will be used by default. It is equivalent to cmpli crfD,0,rA,UIMM.

Count leading zero double word
cntlzd   rA,rS

cntlzd cntlzd.
- **64 bit PowerPC.**

A count of the number of consecutive zero bits of register S is placed into rA. The count is started at bit 0 and the count number can range from 0 to 64, inclusive.

Count leading zero words
cntlzw   rA,rS

cntlzw cntlzw.
• **32 bit PowerPC**

A count of the number of consecutive zero bits of register S is placed into rA. The count is started at bit 0 and the count number can range from 0 to 32, inclusive.

Condition Register AND
crand   crbD,crbA,crbB

crand
• **32 bit PowerPC**

The bit in the condition register specified by crbA is ANDed with the bit in the condition register specified by crbB. The result is placed into the condition register bit specified by crbD.

This instruction can be used to logically reduce the results from several comparisons stored in several condition register fields, prior to branching on the final result.

The MPC601 supports an option with this instruction to use the link register, although this is an illegal instruction in the PowerPC architecture. Although this instruction is supported, it leaves the link register in an undefined state. The best advice is to ignore it!

Condition Register AND with Complement
crandc   crbD,crbA,crbB

crandc
• **32 bit PowerPC**

The bit in the condition register specified by crbA is ANDed with the complement of the bit in the condition register specified by crbB and the result is placed into the condition register bit specified by ubD.

This instruction can be used to logically reduce the results from several comparisons stored in several condition register fields, prior to branching on the final result.

The MPC601 supports an option with this instruction to use the link register, although this is an illegal instruction in the PowerPC architecture. Although this instruction is supported, it leaves the link register in an undefined state. The best advice is to ignore it!

Condition Register Clear
crclr   bitx

crclr
• Alternative mnemomic

This alternative will clear the bit bitx in the condition register. Bitx can be an integer or can be constructed from a group of symbols e.g. 4*cr5 + le will clear the le bit in the cr5 field, eq will clear the eq bit in the cr0 field and so on. The equivalent instruction is crxor bitx,bitx,bitx. The individual bits within the cr fields can use the following symbols:

Symbol	Value	Meaning
lt	0	Less than (bit 0 within a CR field).
gt	1	Greater than (bit 1 within a CR field).
eq	2	Equal to (bit 2 within a CR field).
so	3	Summary overflow (bit 3 within a CR field).
un	3	Unordered(floating point comparison) (bit 2 within a CR field).

Condition Register Equivalent
creqv   crbD,crbA,crbB

creqv
• 32 bit PowerPC

The bit in the condition register specified by crbA is XORed with the bit in the condition register specified by crbB. The complemented result is placed into the condition register bit speciiied by crbD.

This instruction can be used to logically reduce the results from several comparisons stored in several condition register fields, prior to branching on the final result.

The MPC601 supports an option with this instruction to use the link register, although this is an illegal instruction in the PowerPC architecture. Although this instruction is supported, it leaves the link register in an undefined state. The best advice is to ignore it!

Condition Register Move
crmove   bitx, bity

crmove
• Alternative mnemomic

This alternative will move the bit bitx in the condition register to bit y. Bitx and bity can be an integer or can be constructed from a group of symbols e.g. 4*cr5 + le will select the le bit in the cr5 field, eq will select the eq bit in the cr0 field and so on. The equivalent instruction is cror bitx,bity,bity.The individual bits within the cr fields can use the following symbols:

Symbol	Value	Meaning
lt	0	Less than (bit 0 within a CR field).
gt	1	Greater than (bit 1 within a CR field).
eq	2	Equal to (bit 2 within a CR field).
so	3	Summary overflow (bit 3 within a CR field).
un	3	Unordered(floating point comparison) (bit 2 within a CR field).

Condition Register NAND
crnand   crbD,crbA,crbB

crnand
• 32 bit PowerPC

The bit in the condition register specified by crbA is ANDed with the bit in the condition register specified by crbB. The complemented result is placed into the condition register bit specified by crbD.

This instruction can be used to logically reduce the results from several comparisons stored in several condition register fields, prior to branching on the final result.

The MPC601 supports an option with this instruction to use the link register, although this is an illegal instruction in the PowerPC architecture. Although this instruction is supported, it leaves the link register in an undefined state. The best advice is to ignore it!

Condition Register NOR
crnor   crbD,crbA,crbB

crnor
• 32 bit PowerPC

The bit in the condition register specified by crbA is ORed with the bit in the condition register specified by crbB. The complemented result is placed into the condition register bit specified by crbD.

This instruction can be used to logically reduce the results from several comparisons stored in several condition register fields, prior to branching on the final result.

The MPC601 supports an option with this instruction to use the link register, although this is an illegal instruction in the PowerPC architecture. Although this instruction is supported, it leaves the link register in an undefined state. The best advice is to ignore it!

Condition Register Not
crnot bitx, bity
crnot
• Alternative mnemomic

___

This alternative will invert the bit bitx in the condition register and place the result in bit bity. Bitx and bity can be an integer or can be constructed from a group of symbols e.g. 4*cr5 + le will select the le bit in the cr5 field, eq will select the eq bit in the cr0 field and so on. The equivalent instruction is crnor bitx,bity,bity. The individual bits within the cr fields can use the following symbols:

Symbol	Value	Meaning
lt	0	Less than (bit 0 within a CR field).
gt	1	Greater than (bit 1 within a CR field).
eq	2	Equal to (bit 2 within a CR field).
so	3	Summary overflow (bit 3 within a CR field).
un	3	Unordered(floating point comparison) (bit 2 within a CR field).

Condition Register OR
cror crbD,crbA,crbB
cror
• 32 bit PowerPC

___

The bit in the condition register specified by crbA is ORed with the bit in the condition register specified by crbB. The result is placed into the condition register bit specified by crbD.

This instruction can be used to logically reduce the results from several comparisons stored in several condition register fields, prior to branching on the final result.

The MPC601 supports an option with this instruction to use the link register, although this is an illegal instruction in the PowerPC architecture. Although this instruction is supported, it leaves the link register in an undefined state. The best advice is to ignore it!

Condition Register OR with Complement
crorc crbD,crbA,crbB
crorc
• 32 bit PowerPC

___

The bit in the condition register specified by crbA is ORed with the complement of the bit in the condition register specified by crbB and the result is placed into the condition register bit specified by crbD.

Condition Register OR with Complement

crorc   crbD,crbA,crbB

crorc
• **32 bit PowerPC**

The bit in the condition register specified by crbA is ORed with the complement of the bit in the condition register specified by crbB. The result is placed into the condition register bit specified by crbD.

This instruction can be used to logically reduce the results from several comparisons stored in several condition register fields, prior to branching on the final result.

The MPC601 supports an option with this instruction to use the link register, although this is an illegal instruction in the PowerPC architecture. Although this instruction is supported, it leaves the link register in an undefined state. The best advice is to ignore it!

Condition Register Set

crset   bitx

crset
• **Alternative mnemomic**

This alternative will set the bit bitx in the condition register. Bitx can be an integer or can be constructed from a group of symbols e.g. 4*cr5 + le will set the le bit in the cr5 field, eq will set the eq bit in the cr0 field and so on. The equivalent instruction is creqv bitx,bitx,bitx. The individual bits within the cr fields can use the following symbols:

Symbol	Value	Meaning
lt	0	Less than (bit 0 within a CR field).
gt	1	Greater than (bit 1 within a CR field).
eq	2	Equal to (bit 2 within a CR field).
Symbol	**Value**	**Meaning**
so	3	Summary overflow (bit 3 within a CR field).
un	3	Unordered(floating point comparison) (bit 2 within a CR field).

Condition Register XOR
crxor   crbD,crbA,crbB

crxor
• **32 bit PowerPC**

The bit in the condition register specified by crbA is XORed with the bit in the condition register specified by crbB. The result is placed into the condition register bit specified by crbD.

This instruction can be used to logically reduce the results from several comparisons stored in several condition register fields, prior to branching on the final result.

The MPC601 supports an option with this instruction to use the link register, although this is an illegal instruction in the PowerPC architecture. Although this instruction is supported, it leaves the link register in an undefined state. The best advice is to ignore it!

Data Cache Block Flush
dcbf   rA,rB

dcbf
• **32 bit PowerPC**

The EA is the sum (rA|0)+(rB). This instruction is used to force the data cache to flush a block of its data out to memory. The block is the one that is mapped to the effective address. The instruction is used to ensure coherency or to write out a block of data in the cache after manipulating it in the faster cache memory. Graphics and image data can be efficiently processed in this way.

The action taken depends on the state of the block and not the status of the addressed location. The resulting actions occur regardless of whether the page or block is marked as write-through, cache inhibited or cache allowed and depend on the state of the WIM bit and the data block itself.

Coherency Required (WIM = xx1)

Unmodified Block	invalidates all other copies in all processors caches.
Modified Block	writes the block to memory, and invalidates all copies in all processors caches.
Absent Block	causes all other processors to invalidate their copies. If one has a modified copy in cache, this data is written out to memory

Coherency Not Required (WIM = xx0)

Unmodified Block	invalidates the block in the processor's cache.
Modified Block	copies the block to memory. Invalidates the block in the processor's cache.
Absent Block	does nothing.

Note that accessing an invalid address, i.e. one without a valid TLB entry, will not cause a data address exception.

Data Cache Block Invalidate
dcbi   rA,rB

dcbi
• 32 bit PowerPC

The EA is the sum (rA|0)+(rB). This supervisor level instruction is used to invalidate a block of data in the data cache. Unlike dcbf, it does not preserve modified data. Any such information is simply discarded and is not written out to main memory.

The action taken depends on the state of the block and not the status of the addressed location. The resulting actions occur regardless of whether the page or block is marked as write-through, cache inhibited or cache allowed and depend on the state of the WIM bit and the data block itself.

Coherency Required (WIM = xx1)

Unmodified Block	invalidates all other copies in all processors' caches.
Modified Block	invalidates all copies in all processors' caches. The modified data is discarded.
Absent Block	causes all other processors to invalidate their copies. Any modified data is discarded

Coherency Not Required (WIM = xx0)

Unmodified Block	invalidates the block in the local cache.
Modified Block	invalidates the block in the local cache and discards any modified data.
Absent Block	does nothing.

When data address translation is enabled and the logical address has no translation, a data access exception occurs. This instruction is treated as a store to the effective address with respect to protection and address translation and as a result, the reference and change bits will be updated accordingly.

If the effective address accesses an I/O area, i.e. the T bit is set to one in the corresponding segment register, the instruction executes like a nop.

Data Cache Block Store
dcbst   rA,rB

dcbst
• 32 bit PowerPC

The EA is the sum (rA|O)+(rB). This instruction is used to update main memory if a modified copy of data exists in any processor cache block as selected by the effective address. The instruction is similar in action to dcbf except that the block entry is not invalidated. It is useful in ensuring that the main memory contains valid data at the specified effective address. It is usually executed in advance of a program actually needing to access the data.

The action taken depends on the state of the block and not the status of the addressed location and will occur regardless of whether the page or block is marked as write-through, cache inhibited or cache allowed.

If the address requires coherency and any processor cache has modified data, the instruction will cause the modified data to be written out to main memory. If coherency is not required or there is no cached modified data, no further action is taken.

This instruction is treated as a load to the effective address with respect to protection and address translation and as a result, the reference and change bits will be updated accordingly.

If the effective address accesses an I/O area, i.e. the T bit is set to one in the corresponding segment register, the instruction executes like a nop.

Data Cache Block Touch
dcbt   rA,rB

dcbt
• **32 bit PowerPC**

The EA is the sum (rA|0)+(rB). This instruction provides a method for improving performance, as it allows a program to instruct the bus mechanism to preload data into the cache before the program needs to load from it. During the uploading, the program can continue executing in parallel. When it needs to access the data, it can be accessed from the cache rather than from the slow memory. The key is in issuing the dcbt instruction early enough in the program flow to allow enough time for the data to be fetched before the program needs it.

To execute correctly, the effective address must have a hit in MMU TLBs, and have the correct access permission. If the address is an I/O segment or cache inhibited, the access will go ahead but the cache will not be updated. If the address does not hit in the TLBs, the instruction is treated as a nop and does not cause an exception.

Do not assume that it will fetch a complete cache sector o line. The MPC601 will fetch the appropriate 32 byte sector, a addressed by the effective address, but may not fetch the othe sector to complete its cache line length of 64 bytes.

This instruction never affects the reference or change bits in the hashed page table, but its successful completion will change the state of the TLB and cache LRU bits.

Although this instruction executes exactly like the dcbts instruction on the MPC601, it assumes that the program will need to load data from the cache block as opposed to writing to it. A the cache mechanisms are part of the SUPERVISOR model and are implementation and processor dependent, this distinction may become more important than it is at present with the MPC601.

Data Cache Block Touch for Store

dcbtst rA,rB

dcbtst
• 32 bit PowerPC
_____

The EA is the sum $(rA|0)+(rB)$. This instruction provides a method for improving performance, as it allows a program to instruct the bus mechanism to preload data into the cache before the program needs to store data in it. During the uploading, the program can continue executing in parallel. When it needs to store data, it can use the cache rather than from the slow memory. The key is in issuing the dcbtst instruction early enough in the program flow to allow enough time for the data to be fetched before the program needs to store data into the block.

To execute correctly, the effective address must have a hit in MMU TLBs, and have the correct access permission. If the address is an I/O segment or cache inhibited, the access will go ahead but the cache will not be updated. If the address does not hit in the TLBs, the instruction is treated as a nop and does not cause an exception.

Do not assume that it will fetch a complete cache sector or line. The MPC601 will fetch the appropriate 32 byte sector, as addressed by the effective address, but may not fetch the other sector to complete its cache line length of 64 bytes.

This instruction never affects the reference or change bits in the hashed page table, but its successful completion will change the state of the TLB and cache LRU bits.

Although this instruction executes exactly like the dcbt instruction on the MPC601, it assumes that the program will need to store data into the cache block as opposed to load from it. As the cache mechanisms are part of the SUPERVISOR model and are implementation and processor dependent, this distinction may become more important than it is at present with the MPC601.

Data Cache Block Set to Zero

dcbz rA,rB

dcbz
• 32 bit PowerPC
_____

The EA is the sum $(rA|0)+(rB)$. If the block containing the byte addressed by the EA is in the data cache, its bytes are cleared to zero. If the block is not in the data cache and the page is cache allowed, the block is allocated with all bytes cleared to 0. If the address is cache inhibited or write-through, then the system alignment exception handler is invoked. The handler should clear all the bytes in main memory that correspond with the block. If required, the instruction will ensure coherency with any other processors.

This instruction is treated as a store to the effective address with respect to protection and address translation.

If the effective address accesses an I/O area, i.e. the T bit is set to one in the corresponding segment register, the instruction executes like a nop.

This instruction can be used to allocate a cache block in advance without having to incurr memory cycles to fetch the data from memory. It is particularly useful when the program knows that it will store into the cache locations, and so the cache contents initially is of no great importance.

Divide
div   rD,rA,rB
div div. divo divo.
• POWER • MPC601.

This instruction takes a 64 bit value created from the concatenated registers rA and MQ, and divides it by the 32 bit signed contents of rB. The 32 bit signed dividend is placed in rD and the MQ register updated with the 32 bit signed remainder. The division is performed using signed integers. If the quotient cannot be represented in 32 bits, both the SO and OV overflow bits will be set to one, providing the instruction requests such an update by using the o suffix. In such cases, the contents of rD and the MQ register and the corresponding condition register are undefined, except for the special case of $-2^{31} \div -1$. Here, the MQ register is cleared and -2^{31} is put into rD.

Dividing by zero would cause an overflow. It is prudent to perform a bound check on the operands before executing the instruction or alternatively check for overflow afterwards.

The remainder will have the same sign as the dividend. However, a zero remainder is positive.

Divide Double Word
divd   rD,rA,rB
divd divd. divdo divdo.
• 64 bit PowerPC.

This instruction takes the 64 bit signed integer in register A and divides it by the 64 bit signed contents of register B. The 32 bit signed dividend is placed in rD. No remainder is supplied. If needed, this must be caculated by multiplying the quotient by the divisor and subtracting the result from the dividend.

divd rD,rA,rB	; perform the division, store the quotient in rD
mulld rD,rD,rB	; multiply the quotient by the divisor rB and store in rD
subf rD,rD,rA	; subtract rD from the dividend to get the remainder

Divide Double Word Unsigned
divdu   rD,rA,rB

divdu divdu. divduo divduo.
• 64 bit PowerPC.

This instruction takes the 64 bit unsigned integer in register A, and divides it by the 64 bit unsigned contents of register B. The 32 bit unsigned dividend is placed in rD. No remainder is supplied. If needed, this must be caculated by multiplying the quotient by the divisor and subtracting the result from the dividend.

divdu rD,rA,rB	; perform the division, store the quotient in rD
mulld rD,rD,rB	; multiply the quotient by the divisor rB and store in rD
subf rD,rD,rA	; subtract rD from the dividend to get the remainder

Divide Short
divs   rD,rA,rB

divs divs. divso divso.
• POWER • MPC601.

This instruction takes the 32 bit signed integer in register A, and divides it by the 32 bit signed contents of register B. The 32 bit signed dividend is placed in rD and the MQ register updated with the 32 bit signed remainder. The division is performed using signed integers. If the quotient cannot be represented in 32 bits, both the SO and OV overflow bits will be set to one, providing the instruction requests such an update by using the o suffix. In such cases, the contents of rD and the MQ register and the corresponding condition register are undefined, except for the special case of -231 ÷ -1. Here, the MQ register is cleared and -231 is put into rD.

Dividing by zero would cause an overflow. It is prudent to perform a bound check on the operands before executing the instruction or alternatively check for overflow afterwards.

The remainder will have the same sign as the dividend. However, a zero remainder is positive.

Divide Word
divw   rD,rA,rB
divw divw. divwo divwo.
• **32 bit PowerPC**

This instruction takes the 32 bit signed integer in register A, and divides it by the 32 bit signed contents of register B. The 32 bit signed dividend is placed in rD. No remainder is supplied. If needed, this must be caculated by multiplying the quotient by the divisor and subtracting the result from the dividend.

```
divw rD,rA,rB     ; perform the division, store the quotient in
                    rD
mull rD,rD,rB     ; multiply the quotient by the divisor rB and
                    store in rD
subf rD,rD,rA     ; subtract rD from the dividend to get the
                    remainder
```

The division is performed using signed integers. If the quotient cannot be represented in 32 bits, the OV overflow bit will be set to one, providing the instruction requests such an update by using the o suffix. In such cases, the contents of rD and the corresponding condition register are undefined.

Either dividing by zero or the most negative number by -1 would cause an overflow. It is prudent to perform a bound check on the operands before executing the instruction or alternatively check for overflow afterwards.

Divide Word Unsigned
divwu   rD,rA,rB
divwu divwu. divwuo divwuo.
• **32 bit PowerPC**

This instruction takes the 32 bit unsigned integer in register A, and divides it by the 32 bit unsigned contents of register B. The 32 bit unsigned dividend is placed in rD. No remainder is supplied. If needed, this must be caculated by multiplying the quotient by the divisor and subtracting the result from the dividend.

```
divwu rD,rA,rB    ; perform the division, store the quotient in
                    rD
mull  rD,rD,rB    ; multiply the quotient by the divisor rB and
                    store in rD
subf  rD,rD,rA    ; subtract rD from the dividend to get the
                    remainder
```

The division is performed using unsigned integers. If the quotient cannot be represented in 32 bits, as is the case with a divide by zero, the OV overflow bit will be set to one, providing the instruction requests such an update by using the o suffix. In such cases, the contents of rD and the corresponding condition register are undefined.

It is prudent to perform a bound check on the operands before executing the instruction or alternatively check for overflow afterwards.

Difference or Zero
doz   rD,rA,rB

doz doz. dozo dozo.
• POWER • MPC601.

The sum of the NOT(rA)+ (rB) + 1 is placed into register rD. If the value in register rA is algebraically greater than the value in register rB, register rD is cleared.

Difference or Zero Immediate
dozi   rD,rA,SIMM

dozi
• POWER • MPC601.

The sum NOT(rA) + SIMM + 1 is placed into register rD. If the value in register rA is algebraically greater than the value of the SIMM field, register rD is cleared. This instruction is specific to the MPC601.

External Control in Word Indexed
eciwx   rD,rA,rB

eciwx
• 32 bit PowerPC

The effective address is the sum of (rA|0)+(rB). The instruction will attempt to load data using the physical address corresponding to the effective address, with the device specified by the RID bits in the EAR register. If this function is disabled – the E bit in the EAR register is 0 – a data access exception is taken with bit 11 of DSISR set to 1. If the E-bit is set to 1, a load request is generated that bypasses the cache.

If the effective address is mapped to an I/O controller space, i.e. the T bit in the TLB descriptor is set to 1, the instruction is treated as a nop.

This instruction is optional to the PowerPC architecture but is supported on the MPC601. There is no guarantee that all future implementations will support it. In addition, it is likely that the format of the RID bits in the EAR register may also change from implementation to implementation.

External Control out Word Indexed

ecowx   tS,rA,rB

ecowx
• 32 bit PowerPC

The effective address is the sum of (rA|0)+(rB). The instruction will attempt to store data using the physical address corresponding to the effective address, with the device specified by the RID bits in the EAR register. If this function is disabled – the E bit in the EAR register is 0 – a data access exception is taken with bit 11 of DSISR set to 1. If the external access register (EAR) E-bit (bit 0) is set to 1, a load request is generated that bypasses the cache.

If the effective address is mapped to an I/O controller space, i.e. the T bit in the TLB descriptor is set to 1, the instruction is treated as a nop.

This instruction is optional to the PowerPC architecture but is supported on the MPC601. There is no guarantee that all future implementations will support it. In addition, it is likely that the format of the RID bits in the EAR register may also change from implementation to implementation.

Enforce In Order Execution of I/O

eieio

eieio
• 32 bit PowerPC

This instruction is important when accessing I/O areas or other memory addresses where the strict program order must be enforced or completed before continuing the program execution. In short, it will ensure that all currently outstanding accesses to main memory before the execution of the eieio instruction are completed. Any subsequent memory accesses that are requested or initiated will be forced to wait.

Previous accesses are deemed to have completed when they have completed to the point that they cannot cause an exception, all main memory accesses have been completed and the eieio operation has been broadcast onto the bus to comply with cache coherency policy.

This synchronisation will order load and store operations to cache inhibited memory, and store operations to write through cache memory.

This op code simply holds back the program execution to allow all the previous memory requests and operations to complete correctly. For example, if a program has issued several stores to an external peripheral to initialise and request it to perform a function, the eieio instruction can be used to ensure that all the stores have completed in order before reading the peripherals status register. Without, it is possible for the stores not to have completed before the read takes place because of the memory interface and its re-ordering capability, despite the program order.

With the MPC601, eieio performs in the same way as the sync instruction.

Equivalent
eqv   rA,rS,rB

eqv eqv.
• **32 bit PowerPC**

The contents of rS is XORed with the contents of rB and the complemented result is placed into register rA.

Extract and Left Justify Immediate
extlwi   rA,rS,n,b

extlwi extlwi.
• **Alternative mnemomic**

This alternative will extract a field of n bits starting at bit position b in the source register, place the left justified result with all other bits cleared to zero in the target register. The value of n must be greater than 0. It is equivalent of rlwinm rA,rS,b,0,n-1

Extract and Right Justify Immediate
extrwi   rA,rS,n,b

extrwi extrwi.
• **Alternative mnemomic**

This alternative will extract a field of n bits starting at bit position b in the source register, place the right justified result with all other bits cleared to zero in the target register. The value of n must be greater than 0. It is equivalent of rlwinm rA,rS,b+n,32-n,31.

Extend Sign byte
extsb   rA,rS

extsb extsb.
• **32 bit PowerPC**

Register r S[24-31] are placed into rA[24-31]. Bit 24 of rS is placed into rA[0-23].

For example, if rS contains $xxxxxxFF then rA will become $FFFFFFFF, where x can be any value. Similarly, if rS contained $xxxxxx0F then rA would become $0000000F.

Extend Sign Half word
extsh   rA,rS

extsh extsh.
• **32 bit PowerPC**

Register r S[16-31] are placed into rA[16-31]. Bit 16 of rS is placed into rA[0-15].

For example, if rS contains $xxxxFFFF then rA will become $FFFFFFFF, where x can be any value. Similarly, if rS contained $xxxx0xxx then rA would become $00000xxx.

Extend Sign Word
extsw   rA,rS

extsw extsw.
• **64 bit PowerPC.**

Register rS[32-63] are placed into rA[32-63]. Bit 32 of rS is placed into rA[0-31].

For example, if rS contains $xxxxxxxxFFFFFFFF then rA will become $FFFFFFFFFFFFFFFF, where x can be any value. Similarly, if rS contained $xxxxxx0FFFFFFF then rA would become $00000000FFFFFFFF.

Floating Point Absolute
fabs   frD,frB

fabs fabs.
• **32 bit PowerPC**

The contents of frB with bit 0 cleared to 0 is placed into frD.

Floating Point Add
fadd   frD,frA,frB

fadd fadd.
• **32 bit PowerPC**

The floating-point operand in register frA is added to the floating-point operand in register frB. If the most significant bit of the resultant significand is not a one, then the result is normalised. The result is rounded to the target precision under control of the floating-point rounding control field RN of the FPSCR and placed into register frD. Floating-point addition is based on exponent comparison and addition of the two significands.

The floating-point status and control register is updated.

Floating Point Add SIngle Precision

fadds   frD,frA,frB

fadds fadds.

• **32 bit PowerPC**

The floating-point operand in register frA is added to the floating-point operand in register trB. If the most significant bit of the resultant significand is not a one, the result is normalised. The result is rounded to the target precision under control of the floating-point rounding control field RN of the FPSCR and placed into register frD. Floating-point addition is based on exponent comparison and addition of the two significands.

The floating-point status and control register is updated.

Floating Point Convert from Integer Double Word

fcfid   frD,frB

fcfid fcfid.

• **64 bit PowerPC.**

The floating-point operand in register frB is converted to a 64 bit signed integer, using the rounding mode specified by FPSCR[RN], and placed in frD.

If the operand is greater than the largest positive 64 bit integer, then frD will contain 0x7FFF FFFF FFFF FFFF.

If the operand is less than the largest negative 64 bit integer, then frD will contain 0x8000 0000 0000 0000.

Floating Point Compare Ordered

fcmpo   crf D,frA,frB

fcmpo

• **32 bit PowerPC**

The floating-point operand in register frA is compared to the floating-point operand in register frB. The result of the compare is placed into CR field crfD and the FPCC.

If an operand is a NaN, either quiet or signalling, CR field crfD and the FPCC are set to reflect unordered. If an operand is a Signalling NaN, VXSNAN is set, and if invalid operation is disabled (VE=0) then VXVC is set. Otherwise, if an operand is a Quiet NaN, VXVC is set.

Floating Point Compare Unordered

fcmpu   crfD,frA,frB

fcmpu
• **32 bit PowerPC**

The floating-point operand in register frA is compared to the floating-point operand in register frB. The result of the compare is placed into CR field crfD and the FPCC.

If an operand is a NaN, either quiet or signalling, CR field crfD and the FPCC are set to reflect unordered. If an operand is a Signalling NaN, VXSNAN is set.

Floating Point Convert to Integer Double Word

fctid   frD,frB

fctid fctid.
• **32 bit PowerPC**

The floating-point operand in register frB is converted to a 32-bit signed integer, using the rounding mode specified by FPSCR[RN], and placed in bits 32-63 of register trD. Bits 0-31 of register frD are undefined.

Floating Point Convert to Integer DoubleWord with Round toward Zero

fctidz   frD,frB

fctidz fctidz.
• **64 bit PowerPC.**

The floating-point operand in register frB is converted to a 64 bit signed integer, using the rounding mode Round toward Zero, and placed in frD.

If the operand is greater than the largest positive 64 bit integer, then frD will contain 0x7FFF FFFF FFFF FFFF.

If the operand is less than the largest negative 64 bit integer, then frD will contain 0x8000 0000 0000 0000.

Floating Point Convert to Integer Word

fctiw frD,frB

fctiw fctiw.

• 32 bit PowerPC

The floating-point operand in register frB is converted to a 32-bit signed integer, using the rounding mode specified by FPSCR[RN], and placed in bits 32-63 of register trD. Bits 0-31 of register frD are undefined.

Floating Point Convert to Integer Word with Round

fctiwz   frD,frB

fctiwz fctiwz.

• 32 bit PowerPC

The floating-point operand in register frB is converted to a 32-bit signed integer, using the rounding mode Round toward Zero, and placed in bits 32-63 of register frD. Bits 0 to 31 of register frD are undefined.

Floating Point Divide

fdiv   frD,frA,frB

fdiv fdiv.

• 32 bit PowerPC

The floating-point operand in register frA is divided by the floating-point operand in register frB. No remainder is preserved. If the most significant bit of the resultant significand is not a 1, the result is normalised. The result is rounded to the target precision under control of the floating-point rounding control field RN of the FPSCR and placed into register frD. Floating-point division is based on exponent subtraction and division of the significands. FPSCR[FPRF] is set to the class and sign of the result, except for invalid operation exceptions when FPSCR[VE]=1 and zero divide exceptions when FPSCR[ZE]=1.

Floating Point Divide Single Precision

fdivs   frD,frA,frB

fdivs fdivs.

• 32 bit PowerPC

The floating-point operand in register frA is divided by the floating-point operand in register frB. No remainder is preserved.

If the most significant bU of the resultant significand is not a 1, the result is normalised. The result is rounded to the target precision under control of the floating-point rounding control field RN of the FPSCR and placed into register frD. Floating-point division is based on exponent subtraction and division of the significands. FPSCR[FPRF] is set to the class and sign of the result, except for invalid operation exceptions when FPSCR[VE]=1 and zero divide exceptions when FPSCR[ZE]=1.

Floating Point Multiply Add
fmadd   trD,frA,frC,trB

fmadd fmadd.
• **32 bit PowerPC**

The floating-point operand in register frA is multiplied by the floating-point operand in register frC. The floating-point operand in register frB is added to this intermediate result. If the most significant bit of the resultant significand is not a one the result is normalised. The result is rounded to the target precision under control of the floating-point rounding control field RN of the FPSCR and placed into register frD. FPSCR[FPRF] is set to the class and sign of the result, except for invalid operation exceptions when FPSCR[VE]=1.

Floating Point Multiply Add Single Precision
fmadds   frD,frA,frC,frB

fmadds fmadds.
• **32 bit PowerPC**

The floating-point operand in register frA is multiplied by the floating-point operand in register frC. The floating-point operand in registerfrB is added to this intermediate result. If the most significant bit of the resultant significand is not a one the result is normalised. The result is rounded to the target precision under control of the floating-point rounding control field RN of the FPSCR and placed into register frD. FPSCR[FPRF] is set to the class and sign of the result, except for invalid operation exceptions when FPSCR[VE]=1.

Floating Point Register Move
fmr   frD,frB

fmr fmr.
• **32 bit PowerPC**

The contents of register frB is placed into frD.

Floating Point Multiply Subtract

fmsub   frD,frA,frC,frB

fmsub fmsub.
• **32 bit PowerPC**

The floating-point operand in register frA is multiplied by the floating-point operand in register frC. The floating-point operand in register frB is subtracted from this intermediate result. If the most significant bit of the resultant significand is not a one, then the result is normalised. The result is rounded to the target precision under control of the floating-point rounding control field RN of the FPSCR and placed into register frD. FPSCR[FPRF] is set to the class and sign of the result, except for invalid operation exceptions when FPSCR[VE]=1.

Floating Point Multiply Subtract Single Precision

fmsubs   frD,frA,frC,frB

fmsubs fmsubs.
• **32 bit PowerPC**

The floating-point operand in register frA is multiplied by the floating-point operand in register frC. The floating-point operand in register frB is subtracted from this intermediate result. If the most significant bit of the resultant significand is not a one the result is normalised. The result is rounded to the target.precision under control of the tloating-point rounding control field RN of the FPSCR and placed into register frD. FPSCRIFPRF] is set to the class and sign of the result, except for invalid operation exceptions when FPSCR[VE]=1.

Floating Point Multiply

fmul   frD,frA,frC

fmul fmul.
• **32 bit PowerPC**

The floating-point operand in register frA is multiplied by the floating-point operand in register frC. If the most significant bit of the resultant significand is not a one, the result is normalised. The result is rounded to the target precision under control of the floating-point rounding control field RN of the FPSCR and placed into register frD Floating-point multiplication is based on exponent addition and multiplication of the significands FPSCR[FPRF] is set to the class and sign of the result, except for invalid operation exceptions when FPSCR[VE]=1

Floating Point Multiply Single Precision

fmuls   frD,frA,frC

fmuls fmuls.
• **32 bit PowerPC**

The floating-point operand in register frA is multiplied by the floating-point operand in register frC. If the most significant bit of the resultant significand is not a one the result is normalised. The result is rounded to the target precision under control of the floating-point rounding control field RN of the FPSCR and placed into register frD. Floating-point multiplication is based on exponent addition and multiplication of the significands. FPSCR[FPRF] is set to the class and sign of the result, except for invalid operation exceptions when FPSCR[VE]=1.

Floating Point Negate Absolute

fnabs   frD,frB

fnabs fnabs.
• **32 bit PowerPC**

The contents of frB with bit 0 set to one is placed into frD.

Floating Point Negate

fneg   frD,frB

fneg fneg.
• **32 bit PowerPC**

The contents of register frB with bit 0 inverted is placed into register frD.

Floating Point Negative Multiply Add

fnmadd   frD,frA,frC,frB

fnmadd fnmadd.
• **32 bit PowerPC**

The floating-point operand in register frA is multiplied by the floating-point operand in register frC. The floating-point operand in register frB is added to this intermediate result. If the most significant bit of the resultant significand is not a one the result is normalized. The result is rounded to the target precision under control of the floating-point rounding control field RN of the FPSCR, then negated and placed into register frD. This instruction produces the same result as would be obtained by using the floating-point multiply-add instruction and then negating the result, with the following exceptions:

¥   QNaNs propagate with no effect on their sign bit.

¥   QNaNs that are generated as the result of a disabled invalid operation exception have a sign bit of zero.

¥   SNaNs that are converted to QNaNs as the result of a disabled invalid operatiion exception retain the sign bit of the SNaN.

FPSCR[FPRF] is set to the class and sign of the result, except for invalid operation exceptions when FPSCR[VE] = 1.

Floating Point Negative Multiply Add Single Precision
fnmadds   frD,frA,frC,frB

fnmadds fnmadds.

• **32 bit PowerPC**

_____

The floating-point operand in register frA is multiplied by the floating-point operand in register frC. The floating-point operand in register frB is added to this intermediate result. If the most significant bn of the resultant significand is not a one the result is normalised. The result is rounded to the target precision under control of the floating-point rounding control field RN of the FPSCR, then negated and placed into register frD. This instruction produces the same result as would be obtained by using the floating-point multiply-add instruction and then negating the result, with the following exceptions:

¥   QNaNs propagate with no effect on their sign bit.

¥   QNaNs that are generated as the result of a disabled invalid operation exception have a sign bit of zero.

¥   SNaNs that are converted to QNaNs as the result of a disabled invalid operatiion exception retain the sign bit of the SNaN.

FPSCR[FPRF] is set to the class and sign of the result, except for invalid operation exceptions when FPSCR[VE] = 1.

Floating Point Negative Multiply Subtract
fnmsub   trD,frA,frC,frB

nmsub fnmsub.

• **32 bit PowerPC**

_____

The floating-point operand in register frA is multiplied by the floating-point operand in register frC. The floating-point operand in register frB is subtracted from this intermediate result. If the most significant bH of the resultant significand is not a one the result is normalised. The result is rounded to the target precision under control of the floating-point rounding control field RN of the FPSCR, then negated and placed in register frD. This instruction produces the same result as would be obtained by using the floating-point multiply-subtract instruction and then negating the result, with the following exceptions:

¥   QNaNs propagate with no effect on their sign bit.

¥   QNaNs that are generated as the result of a disabled invalid operabon exception have a sign bit of zero.

¥   SNaNs that are converted to QNaNs as the result of a disabled invalid operation exception retain the sign bit of the SNaN.

FPSCR[FPRF] is set to the class and sign of the result, except for invalid operation exceptions when FPSCR[VE]=1.

Floating Point Negative Multiply Subtract Single Precision
fnmsubs   frD,frA,frC,frB

fnmsubs fnmsubs.
• **32 bit PowerPC**

The floating-point operand in register frA is multiplied by the floating point-operand in register frC. The floating-point operand in register frB is subtracted from this intermediate result. If the most significant bit of the resultant significand is not a one the result is normalised. The result is rounded to the target precision under control of the floating-point rounding control field RN of the FPSCR, then negated and placed into register frD. This instruction produces the same result as would be obtained by using the floating-point multiply-subtract instruction and then negating the result, with the following exceptions:

¥   QNaNs propagate with no effect on their sign bit.

¥   QNaNs that are generated as the result of a disabled invalid operation exception have a sign bit of zero.

¥   SNaNs that are converted to QNaNs as the result of a disabled invalid operation exception retain the sign bit of the SNaN.

FPSCR[FPRF] is set to the class and sign of the result, except for invalid operation exceptions when FPSCR[VE]=1.

Floating Point Reciprocal Estimate Single Precision

fres   frD,frB

fres fres.

• **Optional PowerPC(no MPC601 support).**

This instruction will estimate the reciprocal of the value in frB and place it into register frD. This is called an estimate because the precision is only accurate to one part in 256 of the correct result.

Floating Point Round to Single Precision

frsp   frD,frB

frsp frsp.

• **32 bit PowerPC**

If it is already in single-precision range, the floating-point operand in register frB is placed into register frD. Otherwise the floating-point operand in register frB is rounded to single-precision using the rounding mode specified by FPSCR[RN] and placed into register frD. FPSCR[FPRF] is set to the class and sign of the result, except for invalid operation exceptions when FPSCR[VE]=1.

Floating Point Reciprocal Square Root Estimate

frsqrte   frD,frB

frsqrte frsqrte.

• **Optional PowerPC(no MPC601 support).**

A double precision estimate of the reciprocal of the square root of the floating-point value stored in register frB is placed into register frD. The estimate is accurate to 1 part in 32 of the correct value.

Floating Point Select

fsel   frD,frA,frC,frB

fsel fsel.

• **Optional PowerPC(no MPC601 support).**

The floating point operand in frA is compared with zero. If it is greater than or equal to zero, the contents of frC is copied into frD. If it is less than zero or is a NaN, the contents of register frB is copied into frD instead. The comparison treats -0 as being equal to +0.

Floating Point Square Root
fsqrt   frD,frB

fsqrt fsqrt.
• **Optional PowerPC(no MPC601 or MPC603 support).**

The square root of the value contained in frB is placed into register frD. The rounding is under control of the RN control field.

Floating Square Root
fsqrt   frD,frB

fsqrt fsqrt.
• **64 bit PowerPC.**

The square root of the double precision floating point number stored in frB is put into frD. The double precision result will be normalized if the most significant bit of the resultant significand is not one. The result will be rounded using the rounding mode specified by FPSCR[RN].

If the instruction fails due to an invalid operand then an exception is generated and FPSCR[VE] is et to a 1. If the instruction is successful, FPSCR[FPRF] is set to the class and sign of the final result.

Floating Point Square Root Single Precision
fsqrts   frD,frB

fsqrts fsqrts.
• **Optional PowerPC(no MPC601 or MPC603 support).**

The single precision square root of the value contained in frB is placed into register frD. The rounding is under control of the RN control field.

Floating Square Root Single
fsqrts   frD,frB

fsqrts fsqrts.
• **64 bit PowerPC.**

The square root of the single precision floating point number stored in frB is put into frD. The single precision result will be normalized if the most significant bit of the resultant significand is not one. The result will be rounded using the rounding mode specified by FPSCR[RN].

If the instruction fails due to an invalid operand then an exception is generated and FPSCR[VE] is et to a 1. If the instruction is successful, FPSCR[FPRF] is set to the class and sign of the final result.

Floating Point Subtract
fsub   frD,frA,frB

fsub fsub.
• 32 bit PowerPC

The floating-point operand in register trB is subtracted from the floating-point operand in register frA If the most significant bit of the resultant significand is not a one the result is normalised The result is rounded to the target precision under control of the floating-point rounding control field RN of the FPSCR and placed into register frD The execution of the Floating-Point Subtract instruction is identical to that of Floating-Point Add, except that the contents of register frB participates in the operation with its sign bit (bit 0) inverted FPSCR[FPRF] is set to the class and sign of the result, except for invalid operation exceptions when FPSCR[VE]=1

Floating Point Subtract Single Precision
fsubs   frD,frA,frB

fsubs fsubs.
• 32 bit PowerPC

The floating-point operand in register trB is subtracted from the floating-point operand in register frA If the most significant bit of the resultant significand is not a one the result is normalised. The result is rounded to the target precision under control of the floating-point rounding control field RN of the FPSCR and placed into register frD.

Insert from Left Immediate
inslwi   rA,rS,n,b

inslwi inslwi.
• Alternative mnemomic

This alternative will insert a left justified field of n bits into a target register starting at bit position b while leaving all other target register bits unchanged. The value of n must be greater than). It is equivalent of rlwimi rA,rS,32-b,b,(b+n)-1.

Insert from Right Immediate
insrwi   rA,rS,n,b

insrwi insrwi.

• **Alternative mnemomic**

This alternative will insert a right justified field of n bits into a target register starting at bit position b while leaving all other target register bits unchanged. The value of n must be greater than 0. It is equivalent of rlwimi rA,rS,32-(b+n),b,(b+n)-1.

Instruction Synchronise
isync

isync

• **32 bit PowerPC**

This instruction waits for all previous instructions to complete, and then discards any prefetched instructions, causing subsequent instructions to be fetched (or refetched) from memory and to execute in the context established by the previous instructions. This instruction has no effect on other processors or on their caches and is context synchronising.   .

It is often used before and after modifying the segment or BAT registers to ensure that any current memory operation uses the correct descriptor.

Load Address
la   rD,d(rA)

la

• **Alternative mnemomic**

This alternative mnemonic is equivalent to addi rD,rA,d. This is freuqently used to obtain the address of a variable where d(rA) is replaced by a variable's name and the assembler converts this into offset from the base register it has been told to use.

Load Byte and Zero
lbz   rD,d(rA)

lbz

• **32 bit PowerPC**

The effective address is rA indexed by d. The byte in memory addressed by the d(rA) is loaded into register rD[24-31]. The remaining bits (0 to 23) in register rD are cleared to 0.

Load Byte and Zero with Update
lbzu   rD,d(rA)

lbzu
• **32 bit PowerPC**

The effective address (EA) is the sum (rA|O)+d. The byte in memory addressed by the EA is loaded into register rD[24-31]. The remaining bits in register rD are cleared to 0. The EA is placed into register rA.

If operand rA is r0, the MPC601 does not update r0, or if rA=rD, the load data is loaded into register rD and the register update is suppressed. The MPC601 will execute these cases although the PowerPC architecture does not allow them. There is no guarantee that other PowerPC processors will execute this instruction in these cases.

Load Byte and Zero with Update Indexed
lbzux   rD,rA,rB

lbzux
• **32 bit PowerPC**

The effective address (EA)is the sum (rA|0)+(rB). The byte addressed by the EA is loaded into register rD[24~31]. The remaining bits in register rD are cleared to 0. The EA is placed into register rA.

If operand rA is r0, the MPC601 does not update r0, or if rA=rD, the load data is loaded into register rD and the register update is suppressed. The MPC601 will execute these cases although the PowerPC architecture does not allow them. There is no guarantee that other PowerPC processors will execute this instruction in these cases.

Load Byte and Zero Indexed
lbzx   rD,rA,rB

lbzx
• **32 bit PowerPC**

The effective address is the sum (rA|0)+(rB). The byte in memory addressed by the EA is loaded into register rD[24 31] . The remaining bits in register rD are cleared to 0.

Load Double Word
ld   frD,d(rA)

ld

• **64 bit PowerPC.**

The effective address is the sum (rA|0) +(ds || 0b00). The offset ds is concatenated with 0b00 to ensure that it is on a word boundary. The double-word in memory addressed by the EA is then placed into register rD.

Load Double Word and Reserve Indexed
ldarx   rD,rA,rB

ldarx

• **64 bit PowerPC.**

This instruction is used to implement a semaphore or other control structure and is used in conjunction with the stwcx. instruction.

The effective address is the sum (rA|0)+(rB). The word in memory addressed by the EA is loaded into register rD. This instruction creates a reservation for use by a later stwcx. instruction. An address computed from the EA is stored with the reservation, and replaces any address previously associated with the reservation. The EA must be a multiple of 8 unlike its 32 bit version which will accept a value of four. If it is not, the alignment exception handler will be invoked or the results may be undefined.

When the stwcx. is executed, its effective address is compared to the one stored with the reservation. If the reservation is still valid – another bus access has not been snooped by the processor – the stwcx. instruction will perform the store. If not the the store is not performed.

Load Double Word with Update
ldu   frD,rA,rB

ldu

• **64 bit PowerPC.**

The effective address is the sum (rA|0) +(ds || 0b00). The offset ds is concatenated with 0b00 to ensure that it is on a word boundary. The double-word in memory addressed by the EA is then placed into register rD. The EA is then placed into rA thus performing the update.

If rA is r0 or rA = rD, then the instruction is invalid and is not supported by the PowerPC architecture.

Load Double Word with Update Indexed
ldux   frD,rA,rB

ldux

• 64 bit PowerPC.

The effective address is the sum (rA) + (rB). The double-word in memory addressed by the EA is then placed into register rD. The EA is then placed into rA thus performing the update.

If rA is r0 or rA = rD, then the instruction is invalid and is not supported by the PowerPC architecture.

Load Double Word Indexed
ldx   frD,d(rA)

ldx

• 64 bit PowerPC.

The effective address is the sum (rA|0) + (rB). The double-word in memory addressed by the EA is then placed into register rD.

If rA is r0 or rA = rD, then the instruction is invalid and is not supported by the PowerPC architecture.

Load Floating Point Double Precision
lfd   frD,d(rA)

lfd

• 32 bit PowerPC

The effective address is the sum (rA|0)+d. The double-word in memory addressed by the EA is placed into register frD.

Load Floating Point Double Precision with Update
lfdu   frD,d(rA)

lfdu

32 bit PowerPC

The effective address is the sum (rA|0)+d. The double-word in memory addressed by the EA is placed into reglster frD. The EA is placed into the register specified by rA.

If rA is r0 then the MPC601 does not update register r0. This form of the instruction is declared invalid by the PowerPC architecture.

Load Floating Point Double Precision with Update Indexed
lfdux   frD,rA,rB

lfdux
• 32 bit PowerPC

The effective address is the sum (rA|0)+(r B). The double-word in memory addressed by the EA is placed into frD. The EA is then used to update the contents of rA.

If r0 is used within the effective address, the MPC601 will not update the register. This form of the instruction is invalid for the PowerPC architecture but the MPC601 supports it.

Load Floating Point Double Precision with Update Indexed
lfdux   frD,rA,rB

lfdux
• 32 bit PowerPC

The effective address is the sum (rA|0)+(r B). The double-word in memory addressed by the EA is placed into register frD.

If rA is r0 then the MPC601 does not update register r0. This form of the instruction is declared invalid by the PowerPC architecture.

Load Floating Point Double Precision Indexed
lfdx   frD,rA,rB

lfdx
• 32 bit PowerPC

The effective address is the sum (rA|0)+(r B). The double word in memory addressed by the EA is placed into register frD

Load Floating Point Single Precision
lfs   frD,d(rA)

lfs
• 32 bit PowerPC

The effective address is the sum (rA|0)+d. The word in memory addressed by the EA is interpreted as a floating-point single-precision operand. This word is converted to floating-point double-precision format and placed into register frD.

Load Floating Point Single Precision with Update

lfsu   frD,d(rA)

lfsu
• 32 bit PowerPC

The effective address is the sum (rA|0)+d. The word in memory addressed by the EA is interpreted as a floating-point single-precision operand. This word is converted to floating-point double-precision and placed into register frD. The EA is placed into the register specified by rA.

If rA is r0 then the MPC601 does not update register r0. This form of the instruction is declared invalid by the PowerPC architecture, but is supported on the MPC601.

Load Floating Point Single Precision with Update Indexed

lfsux   frD,rA,rB

lfsux
• 32 bit PowerPC

The effective address is the sum (rA|0) + (rB). The word in memory addressed by the EA is interpreted as a floating-point single-precision operand. This word is converted to floating-point double-precision and placed into register frD. The EA is placed into the register specified by rA.

If rA is r0 then the MPC601 does not update register r0. This form of the instruction is declared invalid by the PowerPC architecture.

Load Floating Point Single Precision Indexed

lfsx   frD,rA,rB

lfsx
• 32 bit PowerPC

The effective address is the sum (rA|0)+(rB). The word in memory addressed by the EA is interpreted as a floating-point single-precision operand. This word is converted to floating-point double-precision and placed into register frD.

Load Half Word Algebraic
lha   rD,d(rA)

lha

• 32 bit PowerPC

The effective address is the sum (rA)+d. The half-word in memory addressed by the EA is loaded into register rD[16-31]. The remaining bits in register rD are filled with a copy of bit 0 of the loaded half-word.

Load Half Word Algebraic with Update
lhau   rD,d(rA)

lhau

• 32 bit PowerPC

The effective address is the sum (rA|0)+d. The half-word in memory addressed by the EA is loaded into register rD[16-31] . The remaining bits in register rD are filled with a copy of bit 0 of the loaded half-word. The EA is placed into register rA.

If register rA is r0, the MPC601 does not update register r0, or if rA=rD the load data is loaded into register rD and the register update is suppressed. Although the PowerPC architecture defines load with update instructions with operand rA=0 or rA=rD as invalid forms, the MPC601 allows these cases.

Load Half Word Algebraic with Update Indexed
lhaux   rD,rA,rB

lhaux

• 32 bit PowerPC

The effective address is the sum (rA|0)+(rB). The half-word in memory addressed by the EA is loaded into register rD[16-31]. The remaining bits in register rD are filled with a copy of bit 0 of the loaded half-word. The EA is placed into register rA.

If register rA is r0, the MPC601 does not update register r0, or if rA=rD the load data is loaded into register rD and the register update is suppressed. Although the PowerPC architecture defines load with update instructions with operand rA=0 or rA=rD as invalid forms, the MPC601 allows these cases.

Load Half Word Algebraic Indexed
lhax   rD,rA,rB

lhax
• **32 bit PowerPC**

The effective address is the sum (rA|0)+(rB). The half-word in memory addressed by the EA is loaded into register rD[16-31]. The remaining bits in register rD are filled with a copy of bit 0 of the loaded half-word.

Load Half Word Byte-Reverse Indexed
lhbrx   rD,rA,rB

lhbrx
• **32 bit PowerPC**

The effective address is the sum (rA|0)+(rB). Bits 0-7 of the half-word in memory addressed by the EA are loaded into rD[24-31]. Bits 8-15 of the half-word in memory addressed by the EA are loaded into rD[16-23]. The rest of the bits in rD are cleared to 0. This instruction effectively reverses the two bytes that form the half-word. If the half- word contained the hexadecimal number AB, it would appear in the register as BA.

This instruction runs with the same latency as other load instructions on the MPC601 and MPC603. However, this may not be the case with other PowerPC processors.

This instruction can be used translate between big- and little-endian data structures.

Load Half Word and Zero
lhz   rD,d(rA)

lhz
• **32 bit PowerPC**

The effective address is the sum (rA|0)+d. The half-word in memory addressed by the EA is loaded into register rD[16-31]. The remaining bits in rD are cleared to 0.

Load Half Word and Zero with Update

lhzu   rD,d(rA)

lhzu
• 32 bit PowerPC

The effective address is the sum (rA|0)+d. The half-word in memory addressed by the EA is loaded into register rD[16-31]. The remaining bits in register rD are cleared.

The EA is placed into register rA.

If operand rA=0 the MPC601 does not update register r0, or if rA=rD the load data is loaded into register rD and the register update is suppressed. Although the PowerPC architecture defines load with update instructions with operand rA=0 or rA=rD as invalid forms, the MPC601 allows these cases.

Load Half Word and Zero with Update Indexed

lhzux   rD,rA,rB

lhzux
• 32 bit PowerPC

The effective address is the sum (rA|0)+(rB). The half-word in memory addressed by the EA is loaded into register rD[16-31].The remaining bits in register rD are cleared.

The EA is placed into register rA.

Although the PowerPC architecture defines load with update instructions with operand rA=0 or rA=rD as invalid forms, the MPC601 allows these cases.

Load Half Word and Zero Indexed

lhzx   rD,rA,rB.

lhzx
• 32 bit PowerPC

The effective address is the sum (rA|0)+(rB). The half-word in memory addressed by the EA is loaded into register rD[16-31]. The remaining bits in register rD are cleared.

Load Immediate

li   rD,VALUE

li
• Alternative mnemomic

This alternative mnemonic is equivalent to addi rA,0,value

Load ImmediateSigned
lis rD,VALUE

lis

• **Alternative mnemomic**

This alternative mnemonic is equivalent to addis rA,0,value.

Load Multiple Word
lmw rD,d(rA)

lmw

• **32 bit PowerPC**

The effective address is the sum (rA|0)+d. N consecutive words starting at EA are loaded into GPRs starting with rD and going through to r31, where N = 32-D.

Care must be taken to ensure that the effective address is on a word boundary. If the EA is not, i.e. its address is not a multiple of 4, the alignment exception handler may be invoked if a page boundary is crossed.

If rA is included in the sequence of registers that will be loaded, it is missed out except if it is equal to zero where it is loaded.

It is important to save the register contents prior to executing this instruction. Register corruption after the execution of this instruction is often due to the wrong count value using more registers than anticipated or programmed.

Load String and Compare Byte Indexed
lscbx rD,rA,rB

lscbx lscbx.

• **POWER** • **MPC601.**

This is a string search instruction that uses the XER register to hold not only the number of bytes to be searched but the comparison byte that is used to find a match. Bits 25 to 31 of the XER register contain the byte count while bits 16 to 23 contain the comparison byte.

The EA is the sum (rA|0)+(rB). This is then used as a starting point to load bytes into the general purpose registers starting with rD and working through them as they fill up. If necessary, the filling will wrap around through r0. The search will continue until a matching byte has been found or the number of bytes determined by the count value have been loaded. If a byte match is found, that byte is also loaded, but any remaining bytes are not and this unsused storage is left in an undefined state. Bytes are always loaded left to right in the registers.

When XER[25-31]=0 i.e. the count is zero, the content of rD is not modified.

If a match was found, the count of the number of bytes loaded up to and including the matched byte, is placed in XER[25-31].

It is important to save the register contents prior to executing this instruction. Register corruption after the execution of this instruction is often due to the wrong count value using more registers than anticipated or programmed.

Load String Word Immediate
lswi   rD,rA,NB

lswi
• 32 bit PowerPC

The EA is (rA|0). NB is the number of bytes to be loaded and can be any value from 0 to 31. If NB = 0 then it is assumed that 32 bytes are to loaded. The instruction will proceed to load bytes into the registers starting with rD and proceeding to rD+1 and so on. The register counting will wrap around through r0 if necessary. The bytes are loaded into the registers from left to right. Any partially filled registers will have the unused bytes cleared to 0.

Like the store multiple word instruction, if rA is included as part of destination register set that will receive the bytes, it is skipped. The only exception to this is with r0 which will be filled as it is considered not suitable for addressing.

It is important to save the register contents prior to executing this instruction. Register corruption after the execution of this instruction is often due to the wrong count value using more registers than anticipated or programmed.

Load String Word Indexed
lswx   rD,rA,rB

lswx
• 32 bit PowerPC

The EA is the sum (rA|0)+(rB). Bits 25-31 in the XER register contain the number of bytes to be loaded. If this value is zero, the result is undefined. The instruction will proceed to load bytes into the registers starting with rD and proceeding to rD+1 and so on. The register counting will wrap around through r0 if necessary. The bytes are loaded into the registers from left to right. Any partially filled registers will have the unused bytes cleared to 0.

Like the store multiple word instruction, if rA and rB are included as part of destination register set that will receive the bytes, it is skipped. The only exception to this is with r0 which will be filled as it is considered not suitable for addressing.

It is important to save the register contents prior to executing this instruction. Register corruption after the execution of this instruction is often due to the wrong count value using more registers than anticipated or programmed.

Load Word Algebriac
lwa rD,rA,rB

lwa
• **64 bit PowerPC.**

The effective address is the sum (rA|0) + (ds || 0b00). The word in memory addressed by the EA is then placed into the low order bits of the destination register rD[32-63]. The remaining high order bits of rD are filled with a copy of bit 0. The EA is then placed into rA thus performing the update.

Load Word and Reserve Indexed
lwarx rD,rA,rB

lwarx
• **32 bit PowerPC**

This instruction is used to implement a semaphore or other control structure and is used in conjunction with the stwcx. instruction.

The effective address is the sum (rA|0)+(rB). The word in memory addressed by the EA is loaded into register rD. This instruction creates a reservation for use by a later stwcx. instruction. An address computed from the EA is stored with the reservation, and replaces any address previously associated with the reservation. The EA must be a multiple of 4. If it is not, the alignment exception handler will be invoked if the word loaded crosses a page boundary, or the results may be undefined.

When the stwcx. is executed, its effective address is compared to the one stored with the reservation. If the reservation is still valid – another bus access has not been snooped by the processor – the stwcx. instruction will perform the store. If not the the store is not performed.

Load Word Algebriac with Update Indexed
lwaux rD,rA,rB

lwaux
• **64 bit PowerPC.**

The effective address is the sum (rA) + (rB). The word in memory addressed by the EA is then placed into the low order bits of the destination register rD[32-63]. The remaining high order bits of rD are filled with a copy of bit 0 from the word in memory. The EA is then placed into rA thus performing the update.

If rA is r0 or rA = rD, then the instruction is invalid and is not supported by the PowerPC architecture.

Load Word Algebriac Indexed
lwax   rD,rA,rB

lwax

• **64 bit PowerPC.**

The effective address is the sum (rA|0) + (rB). The word in memory addressed by the EA is then placed into the low order bits of the destination register rD[32-63]. The remaining high order bits of rD are filled with a copy of bit 0 from the word in memory.

Load Word Byte-Reverse Indexed
lwbrx   rD,rA,rB

lwbrx

• **32 bit PowerPC**

The effective address is the sum (rA|0)+(rB). Bits 0-7 of the word in mernory addressed by the EA are loaded into rD[24-31]. Bits 8-15 of the word in memory addressed by the EA are loaded into rD[16-23]. Bits 16-23 of the word in memory addressed by the EA are loaded into rD[8-15]. Bits 24-31 of the word in memory addressed by the EA are loaded into rD[0-7].

This instruction is frequently used to reverse the byte ordering and is used to move data from a little-endian to a big-endian system. If the data was stored as hexadecimal ABCD in memory, it would be loaded into the register as DCBA using this instruction. Please note that the bit ordering within the bytes is not changed.

Load Word and Zero
lwz   rD,d(rA)

lwz

• **32 bit PowerPC**

The effective address is the sum (rA|0)+d. The word in memory addressed bythe EA is loaded into register rD[0-31].

Load Word and Zero with Update

lwzu   rD,d(rA)

lwzu
• 32 bit PowerPC

The effective address is the sum (rA|0)+d. The word in memory addressed by the EA is loaded into register rD. The EA is then placed into register rA to update it.

If the operand rA=0, the MPC601 does not update register r0, or if rA=rD, the load data is loaded into register rD and the register update is suppressed. Although the PowerPC architecture defines load with update instructions with operand rA=0 or rA=rD as invalid forms, the MPC601 allows these cases.

Load Word and Zero with Update Indexed

lwzux   rD,rA,rB

lwzux
• 32 bit PowerPC

The effective address, ie the sum (rA|0)+(rB). The word in memory addressed by the EA is loaded into register rD. The EA is placed into register rA.

If the operand rA=0, the MPC601 does not update register r0, or if rA=rD, the load data is loaded into register rD and the register update is suppressed. Although the PowerPC architecture defines load with update instructions with operand rA=0 or rA=rD as invalid forms, the MPC601 allows these cases.

Load Word and Zero Indexed

lwzx   rD,rA,rB

lwzx
• 32 bit PowerPC

The effective address is the sum (rA|0)+(rB). The word in memory addressed by the EA is loaded into register rD.

Mask Generate

maskg   rA,rS,rB

maskg maskg.
• POWER • MPC601.

This instruction creates a mask which is then placed in rA, ready for future use. Two parameters are used to control the operation: mstart and mstop. Bits 27-31 or rS are used to contain mstart, and bits 27-31 of rB are used for mstop. The resulting mask that is created and placed in rA is created as follows:

- ¥ If mstart < mstop+1 then bits in the range (mstart...mstop) will be ones, all others zero.
- ¥ If mstart > mstop+1 then bits in the range (mstart+1...mstop-1) will be zeros, all others ones.
- ¥ If mstart = mstop=1 then the mask will be all ones.

Mask Insert from Register
maskir   rA,rS,rB

maskir maskir.
• POWER • MPC601.

Register rS is inserted into rA under control of the mask in rB. The mask is created using the maskg instruction.

Move Condition Register Field
mcrf   crfD,crfS

mcrf
• 32 bit PowerPC

The contents of crfS are copied into crtD. No other condition register fields are changed.

Normally, the link bit 31 of this instruction is cleared to zero within the PowerPC architecture. However, the MPC601 will accept a version of this instruction with this bit set. Although invalid by the PowerPC architecture, the MPC601 will execute it, but the link register is left in an undefined state.

Move to Condition Register from FPSCR
mcrfs   crfD,crfS

mcrfs
• 32 bit PowerPC

The contents of FPSCR field specified by operand crfS are copied to the CR field specified by operand crfD. All exception bits copied are cleared to zero in the FPSCR.

Move to Condition Register from XER

mcrxr crfD

mcrxr
• 32 bit PowerPC

The contents of XER[0-3] — the summary overflow, overflow and carry flags — are copied into the condition register field designated by crfD. Bit 3 is reserved but is still carried across. All other fields of the condition register remain unchanged. XER[0-3] is then cleared to 0.

Move from Address Space Register

mfasr rD

mfasr
• Alternative mnemomic

This alternative mnemonic is equivalent to mfspr rD,280 and will move the contents the address space register to rD.

Move from Condition Register

mfcr rD

mfcr
• 32 bit PowerPC

The contents of the condition register are placed into rD. This is the complete 32 bit wide condition code register and should not be confused with the mcrf instruction which deals with one of the eight condition register fields.

Move from Count Register

mfctr rD

mfctr
• Alternative mnemomic

This alternative mnemonic is equivalent to mfspr rD,9 and will move the contents the Count register to rD.

Move from Data Address Register
mfdar rD

mfdar
• **Alternative mnemomic**

This alternative mnemonic is equivalent to mfspr rD,19 and will move the contents the data address register to rD.

Move from DBAT Lower Register
mfdbatl rD, n

mfdbatl
• **Alternative mnemomic**

This alternative mnemonic is equivalent to mfspr rD,537 + (2*n) and will move the contents the lower DBAT register to rD.

Move from DBAT Upper Register
mfdbatu rD, n

mfdbatu
• **Alternative mnemomic**

This alternative mnemonic is equivalent to mfspr rD,536 + (2*n) and will move the contents the upper DBAT register to rD.

Move from Decrementer Register
mfdec rD

mfdec
• **Alternative mnemomic**

This alternative mnemonic is equivalent to mfspr rD,22 and will move the contents the decrementer register to rD.

Move from DSISR Register
mfdsisr rD

mfdsisr
• **Alternative mnemomic**

This alternative mnemonic is equivalent to mfspr rD,18 and will move the contents the DSISR register to rD.

Move from External Access Register

mfear rD

mfear
• **Alternative mnemomic**

This alternative mnemonic is equivalent to mfspr rD,282 and will move the contents the external access register to rD.

Move from FPSCR

mffs frD

mffs mffs.
• **32 bit PowerPC**

The contents of the FPSCR are placed into bits 32-63 of register frD, with bits 0-31 undefined.

In the MPC601 , bits 0-31 of floating-point register frD are actually set to the value hexadecimal FFFFFFFF or FFF80000.

Move from IBAT Lower Register

mfibatl rD, n

mfibatl
• **Alternative mnemomic**

This alternative mnemonic is equivalent to mfspr rD,529 + (2*n) and will move the contents the lower IBAT register to rD.

Move from IBAT Upper Register

mfibatu rD, n

mfibatu
• **Alternative mnemomic**

This alternative mnemonic is equivalent to mfspr rD,528 + (2*n) and will move the contents the upper IBAT register to rD.

Move from Link Register

mflr rD

mflr
Alternative mnemomic

This alternative mnemonic is equivalent to mfspr rD,8 and will move the contents of the Link register to rD.

Move from Machine State Register

mfmsr   rD

mfmsr

• **32 bit PowerPC**

The contents of the MSR are placed into rD. This is a supervisor-level instruction.

Move from Processor Version Register

mfpvr   rD

mfpvr
• **Alternative mnemomic**

This alternative mnemonic is equivalent to mfspr rD,287 and will move the contents the processor version register to rD.

Move from SDR1 Register

mfsdr1   rD

mfsdr1
• **Alternative mnemomic**

This alternative mnemonic is equivalent to mfspr rD,25 and will move the contents the SDR1 register to rD.

Move from Special Purpose Register

mfspr   rD,SPR

mfspr
• **32 bit PowerPC**

The SPR field denotes a special purpose register. The encoding is processor specific. If a coding is used that is not supported by the processor, the instruction is treated as invalid. The content of the designated SPR are placed into rD.

For mtspr and mfspr instructions, the SPR number coded in assembly language does not appear directly as a l0-bit binary number in the instruction. The number coded is split into two 5 bit halves that are reversed in the instruction, with the high-order 5 bits appearing in bits 16-20 of the instruction and the low-order 5 bits in bits 11-15.

Move from Special Purpose Register n

mfsprg rD, n

mfsprg
• Alternative mnemomic

This alternative mnemonic is equivalent to mfspr rD,272 + n and will move the contents the SPR n to rD.

Move from Status Save Restore 0 Register

mfsrr0 rD

mfsrr0
• Alternative mnemomic

This alternative mnemonic is equivalent to mfspr rD,26 and will move the contents the SRR0 register to rD.

Move from Status Save Restore 1 Register

mfsrr1 rD

mfsrr1
• Alternative mnemomic

This alternative mnemonic is equivalent to mfspr rD,27 and will move the contents the SRR1 register to rD.

Move from Time Base

mftb rD,TBR

mftb
Optional PowerPC(no MPC601 or MPC603 support).

This instruction move data from either of the time base registers TB or TBU depending on the value of TBR and places the data into the register rD. If TBR= 268, then the register TB is accessed. If TBR=269, then the register TBU is accessed. If the instruction is executed on a 64 bit machine, the high order bits of rD are set to zero.

Move from Time Base Lower Register

mftbl   rD

mftbl
• **Alternative mnemomic**

This alternative mnemonic is equivalent to mfspr rD,268 and will move the contents the lower time base register to rD. Note that the SPR number is different from the mttbl altenative instruction. Reads are not supervisor protected while writes are.

Move from Time Base Upper Register

mftbu   rD

mftbu
• **Alternative mnemomic**

This alternative mnemonic is equivalent to mfspr rD,269 and will move the contents the upper time base register to rD. Note that the SPR number is different from the mttbu altenative instruction. Reads are not supervisor protected while writes are.

Move from XER

mfxer   rD

mfxer
• **Alternative mnemomic**

This alternative mnemonic is equivalent to mfspr rF,1 and will move the contents of the XER register to rD.

Move Register

mr   rA,rS

mr
• **Alternative mnemomic**

This alternative mnemonic is equivalent to or rA,rS,rS. It i used to move the contents of a register from to another register

Move to Address Space Register

mtasr   rS

mtasr
- **Alternative mnemomic**

This alternative mnemonic is equivalent to mtspr 280,rS and will move the contents of rS to the address space register.

Move to Condition Register Fields

mtcrf   CRM,rS

mtcrf
- **32 bit PowerPC**

The contents of rS are placed into the condition register under control of the field mask specified by operand CRM. This is an 8 bit mask contained in bits 12 to 19 of the instruction, where each bit signifies whether the associated condition register field is updated. If bit 12 in the instruction – bit 0 in the CRM – is set to one, the first condition register field will be updated and so on. An example is shown below.

```
Initial CR    0000 0000 0000 0000 0000 0000 0000 0000
      CRM     0    0    1    1    1    0    1    0
      rS      1011 1101 1011 1001 1111 1111 0101 1111
Final   CR    0000 0000 1011 1001 1111 0000 0101 0000
```

Move to Count Register

mtctr   rS

mtctr
- **Alternative mnemomic**

This alternative mnemonic is equivalent to mtspr 9,rS and will move the contents of rS to the Count register.

Move to Data Address Register

mtdar   rS

mtdar
- **Alternative mnemomic**

This alternative mnemonic is equivalent to mtspr 19,rS and will move the contents of rS to the Data address register.

Move to DBAT Lower register
mtdbatl n,rS

mtdbatl
• Alternative mnemomic

This alternative mnemonic is equivalent to mtspr 537 +(2*n),rS and will move the contents of rS to the lower IBAT register n.

Move to DBAT Upper register
mtdbatu n,rS

mtdbatu
• Alternative mnemomic

This alternative mnemonic is equivalent to mtspr 536 +(2*n),rS and will move the contents of rS to the upper DBAT register n.

Move to Decrementer Register
mtdec rS

mtdec
• Alternative mnemomic

This alternative mnemonic is equivalent to mtspr 22,rS and will move the contents of rS to the decrementer register.

Move to DSISR
mtdsisr rS

mtdsisr
• Alternative mnemomic

This alternative mnemonic is equivalent to mtspr 18,rS and will move the contents of rS to the DSISR register.

Move to External Address Register
mtear rS

mtear
• Alternative mnemomic

This alternative mnemonic is equivalent to mtspr 282,rS and will move the contents of rS to the external address register.

Move to FPSCR Bit 0
mtfsb0   crbD

mtfsb0 mtfsb0.
• **32 bit PowerPC**

The bit of the FPSCR specified by operand crbD is cleared to 0. This is used in handling floating exceptions and software assist.

Please note that bits 1 and 2 (FEX and VX) cannot be explicitly reset.

Move to FPSCR Bit 1
mtfsb1   crbD

mtfsb1 mtfsb1.
• **32 bit PowerPC**

The bit of the FPSCR specified by operand crbD is set to 1. This is used in handling floating exceptions and software assist.

Please note that bits 1 and 2 (FEX and VX) cannot be explicitly reset.

Move to FPSCR Fields
mtfsf   FM,frB

mtfsf mtfsf.
• **32 bit PowerPC**

Bits 32-63 of register frB are placed into the condition register under control of the field mask specified by operand FM. This is an 8 bit mask contained in bits 7 to 14 of the instruction, where each bit signifies whether the associated condition register field is updated. If bit 7 in the instruction – bit 0 in the FM – is set to one, the first condition register field will be updated and so on. An example is shown below.

```
Initial FPSCR  0000 0000 0000 0000 0000 0000 0000 0000
          FM    0    0    1    1    1    0    1    0
   frB[32-63]  1011 1101 1011 1001 1111 1111 0101 1111
Final FPSCR    0000 0000 1011 1001 1111 0000 0101 0000
```

Please note that when moving to bits 0 and 3 of the FPSCR that the bits are set using the data from frB. The usual rule of setting the FX bit 0 when an exception bit changes is effectively ignored. Similarly bits 1 and 2 are only cleared if the logical or of all their associated bits and enables are zero.

Move to FPSCR Field Immediate
mtfsfi crfD,IMM
mtfsfi mtfsfi.
• **32 bit PowerPC**

The value of the IMM field is placed into FPSCR field crfD. All other FPSCR fields are unchanged.

Please note that when moving to bits 0 and 3 of the FPSCR that the bits are set using the data from the immediate value IMM. The usual rule of setting the FX bit 0 when an exception bit changes is effectively ignored. Similarly bits 1 and 2 are only cleared if the logical or of all their associated bits and enables are zero.

Move to IBAT Lower register
mtibatl n,rS
mtibatl
• **Alternative mnemomic**

This alternative mnemonic is equivalent to mtspr 529 +(2*n),rS and will move the contents of rS to the lower IBAT register n.

Move to IBAT Upper register
mtibatu n,rS
mtibatu
• **Alternative mnemomic**

This alternative mnemonic is equivalent to mtspr 528 +(2*n),rS and will move the contents of rS to the upper IBAT register n.

Move to Link Register
mtlr rS
mtlr
• **Alternative mnemomic**

This alternative mnemonic is equivalent to mtspr 8,rS and will move the contents of rS to the Link register.

Move to Machine State Register

mtmsr   rS

mtmsr
• **32 bit PowerPC**

The contents of rS are placed into the MSR. This instruction is a supervisor-level instruction and is context synchronising.

Move to SDR1

mtsdr1   rS

mtsdr1
• **Alternative mnemomic**

This alternative mnemonic is equivalent to mtspr 25,rS and will move the contents of rS to the SDR1 register.

Move to Special Purpose Register

mtspr   SPR,rS

mtspr
• **32 bit PowerPC**

The SPR field denotes a special purpose register. The encoding is processor specific. If a coding is used that is not supported by the processor, the instruction is treated as invalid. The contents of rD are placed into the designated SPR.

For mtspr and mfspr instructions, the SPR number coded in assembly language does not appear directly as a l0-bit binary number in the instruction. The number coded is split into two S-bit halves that are reversed in the instruction, with the high-order 5 bits appearing in bits 16-20 of the instruction and the low-order 5 bits in bits 11-15.

Move to Special Purpose Register n

mtsprg   n,rS

mtsprg
• **Alternative mnemomic**

This alternative mnemonic is equivalent to mtspr 272 + n,rS and will move the contents of rS to the special purpose register n.

Move to Status Save Restore Register 0
mtsrr0 rS

mtsrr0
• **Alternative mnemomic**

This alternative mnemonic is equivalent to mtspr 26,rS and will move the contents of rS to the SRR0 register.

Move to Status Save Restore Register 1
mtsrr1 rS

mtsrr1
• **Alternative mnemomic**

This alternative mnemonic is equivalent to mtspr 27,rS and will move the contents of rS to the SRR1 register.

Move to Time Base Lower register
mttbl rS

mttbl
• **Alternative mnemomic**

This alternative mnemonic is equivalent to mtspr 268,rS and will move the contents of rS to the lower time base register.

Move to Time Base Upper register
mttbu rS

mttbu
• **Alternative mnemomic**

This alternative mnemonic is equivalent to mtspr 285,rS and will move the contents of rS to the upper time base register.

Move to XER
mtxer rS

mtxer
• **Alternative mnemomic**

This alternative mnemonic is equivalent to mtspr 1,rS and will move the contents of rS to the XER register.

Multiply
mul   rD,rA,rB
mul mul. mulo mulo.
• **POWER** • **MPC601.**

Bits 0-31 of the product rA*rB are stored in rD and bits 32 to 63 are stored in the MQ register. Although this seems to be a more powerful instruction compared to the normal PowerPC multiply instructions, it does use the MQ register to store the upper product bits and thus its use is restricted to the MPC601 only.

Multiply High Double Word
mulhd   rD,rA,rB
mulhd mulhd.
• **64 bit PowerPC.**

The contents of rA and rB are interpreted as 64-bit signed integers. An 128 bit product is formed of which the high order 64 bits are placed in rD.

This instruction may execute faster if rB contains the operand having the smaller absolute value although any such advantages are processor implementation dependent.

Multiply High Double Word Unsigned
mulhdu   rD,rA,rB
mulhdu mulhdu.
• **64 bit PowerPC.**

The contents of rA and rB are interpreted as 64-bit unsigned integers. An 128 bit product is formed of which the high order 64 bits are placed in rD.

This instruction may execute faster if rB contains the operand having the smaller absolute value although any such advantages are processor implementation dependent.

Multiply High Word
mulhw   rD,rA,rB
mulhw mulhw.
• **32 bit PowerPC**

The contents of rA and rB are interpreted as 32-bit signed integers. The 64 bit product is formed. The high-order 32 bits of the 64-bit product are placed into rD. Both operands and the product are interpreted as signed integers. This instruction may execute faster if rB contains the operand having the smaller absolute value although any such advantages are processor implementation dependent.

Multiply High Word Unsigned
mulhwu   rD,rA,rB

mulhwu mulhwu.
• **32 bit PowerPC**

The contents of rA and of rB are extracted and interpreted as 32-bit unsigned integers. The 64-bit product is formed. The high-order 32 bits of the 64-bit product are placed into rD. Both operands and the product are interpreted as unsigned integers. This instruction may execute faster if rB contains the operand having the smaller absolute value although any such advantages are processor implementation dependent.

With the MPC601, this instruction leaves the state of the MQ register as undefined.

Multiply Low Double Word
mulld   rD,rA,rB

mulld mulld. mulldo mulldo.
• **64 bit PowerPC.**

The contents of rA and rB are interpreted as 64-bit signed integers. An 128 bit product is formed of which the low order 64 bits are placed in rD. If the OE bit is set to 1 then the product cannot be represented in 64 bits.

This instruction may execute faster if rB contains the operand having the smaller absolute value although any such advantages are processor implementation dependent.

Multiply Low Immediate
mulli   rD,rA,SIMM

mulli
• **32 bit PowerPC**

The low-order 32 bits of the 48-bit product (rA)*SIMM are placed into register rD. The low-order 32 bits of the product are independent of whether the operands are treated as signed or unsigned integers, and are the correct 32-bit product. The low-order bits are independent of whether the operands are treated as signed or unsigned integers. However, XER[OV] is set based on the result interpreted as a signed integer. The high-order bits are lost. This instruction can be used with mulhwx to calculate a full 64-bit product.

Multiply Low
mullw   rD,rA,rB

mullw mullw. mullwo mullwo.
• 32 bit PowerPC

The low-order 32 bits of the 64-bit product (rA)*(rB) are placed into register rD. The low-order 32 bits of the product are independent of whether the operands are treated as signed or unsigned integers, and are the correct 32-bit product. However, XER[OV] is set based on the result interpreted as a signed integer. The high-order bits are lost. This instruction can be used with mulhwx to calculate a full 64-bit product. Some implementations may execute faster if rB contains the operand having the smaller absolute.

Negative absolute
nabs   rD,rA

nabs nabs. nabso nabso.
• POWER • MPC601.

The negative absolute value -|(rA)| is placed into register rD.
Please note that this instruction never overflows. If the instruction is overflow enabled, then XER[OV] is cleared to zero and XER[SO] is not changed.

NAND
nand   rA,rS,rB

nand nand.
• 32 bit PowerPC

The contents of rS is ANDed with the contents of rB and the one's complement of the result is placed into register rA. If rA=rB, the NAND instruction can be used to obtain the one's complement.

```
rS    0101 1010 0011 1111 0000 0100 1011 0001
rB    1111 0000 1111 0101 0000 1011 1011 0001
AND   0101 0000 0011 0101 0000 0000 1011 0001
rA    1010 1111 1100 1010 1111 1111 0100 1110
```

The NAND instruction can be used to obtain the one's complement by making rB=rA and making the contents of rS either all ones or the same as rB.

```
rS    0101 1010 0011 1111 0000 0100 1011 0001
rB    0101 1010 0011 1111 0000 0100 1011 0001
AND   0101 1010 0011 1111 0000 0100 1011 0001
rB    1010 0101 1100 0000 1111 1011 0100 1110
```

Negate
neg   rD,rA
neg neg. nego nego.
• **32 bit PowerPC**

The sum -(rA) + 1 is placed into register rD. With two's complement arithmetic, one is added to the number and the sign bit reversed.

The most neagtive number is a special case: If rA already contains it i.e. $8000 0000 then the result placed in rD is the same value and providing the OE bit is set to one, OV will be set.

No-op
nop
nop
• **Alternative mnemomic**

This alternative mnemonic is equivalent to ori 0,0,0 and is the preferred instruction for performing the no-op.

NOR
nor   rA,rS,rB
nor nor.
• **32 bit PowerPC**

The contents of rS is ORed with the contents of rB and the one's complement of the result is placed into register rA.

rS	0101 1010 0011 1111 0000 0100 1011 0001
rB	1111 0000 1111 0101 0000 1011 1011 0001
OR	1111 1010 1111 1111 0000 1111 1011 0001
rA	0000 0101 0000 0000 1111 0000 0100 1110

Complement Register
not   rA,rS
not
• **Alternative mnemomic**

This alternative mnemonic is equivalent to nor rA,rS,rS. It is used to complement a register's contents and move the result to another register.

OR
or   rA,rS,rB

or or.
• 32 bit PowerPC

The contents of rS is ORed with the contents of rB and the result is placed into rA.

```
rS   0101 1010 0011 1111 0000 0100 1011 0001
rB   1111 0000 1111 0101 0000 1011 1011 0001
OR   1111 1010 1111 1111 0000 1111 1011 0001
rA   1111 1010 1111 1111 0000 1111 1011 0001
```

OR with complement
orc   rA,rS,rB

orc orc.
• 32 bit PowerPC

The contents of rS is ORed with the complement of the contents of rB and the result is placed into rA.

```
rS   0101 1010 0011 1111 0000 0100 1011 0001
rB   0000 1111 0000 1010 1111 0100 0100 1110
comp 1111 0000 1111 0101 0000 1011 1011 0001
OR   1111 1010 1111 1111 0000 1111 1011 0001
rA   1111 1010 1111 1111 0000 1111 1011 0001
```

OR Immediate
ori   rA,rS,UIMM

ori
• 32 bit PowerPC

The contents of rS is ORed with x'0000' 11 UIMM and the result is placed into rA. The preferred no-op is ori 0,0,0.

```
rS   0101 1010 0011 1111 0000 0100 1011 0001
UIMM 0000 0000 0000 0000 1111 0000 1111 0101
OR   0101 1010 0011 1111 1111 0100 1111 0101
rA   0101 1010 0011 1111 1111 0100 1111 0101
```

OR Immediate Shifted
oris   rA,rS,UIMM

oris
• 32 bit PowerPC

The contents of rS is ORed with UIMM ‖ $0000 and the result is placed into rA.

```
rS     0101 1010 0011 1111 0000 0100 1011 0001
UIMM   1111 0000 1111 0101 0000 0000 0000 0000
OR     1111 1010 1111 1111 0000 0100 1011 0001
rA     1111 1010 1111 1111 0000 1111 1011 0001
```

Return from Interrupt
rfi —

rfi
• 32 bit PowerPC

Bits 16 to 31 of SRR1 are placed into bits 16-31 of the MSR, then the next instruction is fetched, under control of the new MSR value, using the address stored in register SRR0. This instruction is a supervisor-level instruction and is context synchronising .

This instruction is used to return from an exception – not just an interrupt! – and should be executed at the end of the exception handler.

Rotate Left Double Word then Clear Left
rldcl   rA,rS,rB,MB

rldcl rldcl.
• 64 bit PowerPC.

The contents of rS is rotated left the number of positions specified in the low order six bits of rB. The rotated data is ANDed with mask and inserted into rA. The mask is constructed using 64 bits with bits MB to bit 63 set to 1 and all other bits cleared to zero .This instruction can be used to extract or rotate a bit field as follows:

To obtain a bit field of size x bits starting at bit y, then set the six low order bits of rB to x+y and make MB = 64-x. The result will be placed right justified into rA.

To rotate a register's contents left by x bits set the six low order bits of rB to x and MB = 0.

To rotate a register's contents right by x bits set the six low order bits of rB to 64 - x and MB = 0.

Rotate Left Double Word then Clear Right

rldcr rA,rS,rB,ME

rldcr rldcr.
• 64 bit PowerPC.

The contents of rS is rotated left the number of positions specified in the low order six bits of rB. The rotated data is ANDed with the mask and inserted into rA. The mask is constructed using 64 bits with bits 0 to ME set to 1 and all other bits cleared to zero

This instruction can be used to extract or rotate a bit field as follows:

To obtain a bit field of size x bits starting at bit y, then set the six low order bits of rB to y and make ME= x -1. The result will be placed left justified into rA.

To rotate a register's contents left by x bits set the six low order bits of rB to x and ME = 63.

To rotate a register's contents right by x bits set the six low order bits of rB to 64 - x and ME = 63.

Rotate Left Double Word Immediate then Clear

rldic rA,rS,rB,SH,MB

rldic rldic.
64 bit PowerPC.

The contents of rS is rotated left the number of bits specified by SH. The rotated data is ANDed with the generated mask and inserted into rA. The mask is created by setting all bits from MB through to bit 63-SH to 1 and clearing all other bits to zero.

To clear the high order x bits of a register, set MB = x and SH to 0.

Rotate Left Double Word Immediate then Clear Left

rldicl rA,rS,rB,SH,MB

rldicl rldicl.
64 bit PowerPC.

The contents of rS is rotated left the number of bits specified SH. The rotated data is ANDed with the generated mask and inserted into rA. The mask is created by setting all bits from MB through to bit 63 to 1 and clearing all other bits to zero.

Rotate Left Double Word Immediate then Clear Right
rldicr rA,rS,rB,SH,ME

rldicr rldicr.
• **64 bit PowerPC.**

The contents of rS is rotated left the number of positions specified by SH. The rotated data is ANDed with the generated mask and inserted into rA. The mask is created by setting all bits from bit 0 through to ME to 1 and clearing all other bits to zero.

Rotate Left Double Word Immediate then Mask Insert
rldimi rA,rS,rB,SH,MB

rldimi rldimi.
• **64 bit PowerPC.**

The contents of rS is rotated left the number of positions specified by SH. The rotated data is inserted into rA under control of the genertaed mask. The mask is created by setting all bits from MB through to bit 63 - SH to 1 and clearing all other bits to zero.

Rotate Left then Mask Insert
rlmi rA,rS,rB,MB,ME

rlmi rlmi.
• **POWER** • **MPC601.**

The contents of rS is rotated left the number of positions specified by bits 27-31 of rB. The rotated data is inserted into rA under control of the generated mask. MB is the bit number that is the start of the mask and ME is the corresponding bit number for the end.

Example masks:

```
    MB=4, ME=21, mask = 0000 1111 1111 1111 1111 1100 00
0000
    MB=9, ME=31, mask = 0000 0000 0111 1111 1111 1111 11
1111
```

Rotate Left Word Immediate then Mask insert
rlwimi rA,rS,SH,MB,ME
rlwimi rlwimi.
• 32 bit PowerPC

___

The contents of rS are rotated left by the number of bits given by SH. The rotated data is inserted into rA under control of a mask. This mask is created by setting the bits from the bit specified by MB through the bit specified by ME to one and clearing and all the other bits to zero.

Example masks:
```
MB=4, ME=21, mask = 0000 1111 1111 1111 1111 1100 0000
0000
MB=9, ME=31, mask = 0000 0000 0111 1111 1111 1111 1111
1111
```

Rotate Left Word Immediate then AND with Mask
rlwinm rA,rS,SH,MB,ME
rlwinm rlwinm.
• 32 bit PowerPC

___

The contents of register rS are rotated left by the number of bits specified by operand SH. A mask is generated having 1-bits from the bit specified by operand MB through the bit specified by operand ME and 0-bits elsewhere. The rotated data is ANDed with the generated mask and the result is placed into register rA. This instruction can be used to the extraction, clearing and shifting of bit fields as follows:

¥ Right justified extraction of y bit field starting at position x in rS:

MB=32-y, ME=31, SH=y+x

¥ Left justified extraction of y bit field starting at position x in rS:

MB= y-1, ME=31n-1, SH=b

¥ Rotate left contents of rS by y bits:

MB=0, ME=y-1, SH=y

¥ Rotate right contents of rS by y bits:

MB=0, ME=y-1, SH=31-y

¥ Shift right the contents of rS by y bits:

MB=y, ME=31, SH=32-y

¥   Clear high order y bits of rS and then shift the result left by x bits:

MB=y-x, ME=31-x, SH=x

¥   Clear low order y bits of rS:

MB=0, ME=31-y, SH=0

Example masks:
```
MB=4, ME=21, mask = 0000 1111 1111 1111 1111 1100 0000
0000
    MB=9, ME=31, mask = 0000 0000 0111 1111 1111 1111 1111
1111
```

Rotate Left Word then AND with Mask
rlwnm   rA,rS,rB,MB,ME
rlwnm rlwnm.
• 32 bit PowerPC

The contents of rS are rotated left by the number of bits given by the number stored in bits 27 to 31 of rB. The rotated data is ANDed with the generated mask and the result inserted into rA under control of the mask. This mask is created by setting the bits from the bit specified by MB through the bit specified by ME to one and clearing and all the other bits to zero.

Example masks:
```
MB=4, ME=21, mask = 0000 1111 1111 1111 1111 1100 0000
0000
    MB=9, ME=31, mask = 0000 0000 0111 1111 1111 1111 111
1111
```

Rotate Left Immediate
rotlwi   rA,rS,n
rotlwi rotlwi.
• Alternative mnemomic

This alternative will rotate the contents of a register left by n bits without any masking operation.. It is equivalent of rlwin rA,rS,n,0,31.

Rotate Right Immediate
rotrwi   rA,rS,n

rotrwi rotrwi.
• Alternative mnemomic

This alternative will rotate the contents of a register right by n bits without any masking operation.. It is equivalent of rlwinm rA,rS,32-n,0,31.

Rotate Left
rotrwi   rA,rS,rB

rotrwi rotrwi.
• Alternative mnemomic

This alternative will rotate the contents of a register left by the number of bits specified in rB without any masking operation.. It is equivalent of rlwnm rA,rS,rB,0,31.

Rotate Right and Insert Bit
rib   rA,rS,rB

rib rrib.
POWER • MPC601.

Bit 0 of rS is rotated right the amount specified by bits 27-31 of rB. The bit is then inserted into rA.

The PowerPC architecture equivalent is rlwinm.

System Call
sc

c

32 bit PowerPC

When executed, the effective address of the instruction following the sc instruction is placed into SRR0. Bits 16-31 of the MSR are placed into bits 16-31 of SRR1, and bits 0-15 of SRR1 are set to undefined values. Then a system call exception is generated. The exception causes the next instruction to be fetched from offset $C000 from the base physical address indicated by the new setting of MSR[IP]. This location should contain the system call exception handler.

To ensure compatibility with future versions of the PowerPC architecture, bits 16-29 should be coded as zero and bit 30 should be coded as a 1. This instruction is context synchronising.

SLB Invalidate All
slbia

slbia
• 64 bit PowerPC.

The complete segment lookaside buffer (SLB) is invalidated irrespective of the state of MSR[IR] and MSR[DR].

This is a supervisor level instruction and is optional within the PowerPC architecture.

SLB Invalidate Entry
slbie   rB

slbie
• 64 bit PowerPC.

The entry corresponding to the effective address obtained from the contents of rB is invalidated if present within the segment lookaside buffer (SLB) irrespective of the state of MSR[IR] and MSR[DR]. This is a supervisor level instruction and is optional within the PowerPC architecture.

Shift Left Double Word
sld   rA,rS,rB

sld sld.
• 64 bit PowerPC.

Register rS is shifted left n bits where n is the shift amount specified by the seven low order bits of rB. The vacated positions are filled with zeros. The result is then put into rA. If the shift value within rB is between 64 and 127, the result is zero(i.e. all the bits have been shifted and filled with zeros).

Shift Left Extended
sle   rA,rS,rB

sle sle.
• POWER • MPC601.

Register rS is rotated left n bits where n is the shift amount specified in bits 27-31 of register rB. The rotated word is placed in the MQ register. A mask of 32-n ones followed by n zeros is generated. The logical AND of the rotated word and the generated mask is placed in register rA.

Shift Left Extended with MQ

sleq rA,rS,rB

sleq sleq.
• POWER • MPC601.

Register rS is rotated left n bits where n is the shift amount specified in bits 27-31 of register rB. A mask of 32-n ones followed by n zeros is generated. The rotated word is then merged with the contents of the MQ register, under control of the generated mask. The merged word is put into the MQ register.

Shift Left immediate with MQ

sliq rA,rS,SH

sliq sliq.
• POWER • MPC601.

Register rS is rotated left n bits where n is the shift amount specified by operand SH. The rotated word is placed in the MQ register. A mask of 32-n ones followed by n zeros is generated. The logical AND of the rotated word and the generated mask is placed into register rA.

Shift Left Long Immediate with MQ

slliq rA,rS,SH

slliq slliq.
• POWER • MPC601.

Register rS is rotated left n bits where n is the shift amount specified by SH. A mask of 32-n ones followed by n zeros is generated. The rotated word is then merged with the contents of MQ, under control of the generated mask. The merged word is placed into rA. The rotated word is placed into the MQ register.

Shift Left Long with MQ

sllq rA,rS,rB

sllq sllq.
• POWER • MPC601.

Register rS is rotated left n bits where n is the shift amount specified in bits 27-31 of register rB. When bit 26 of register rB is a zero, a mask of 32-n ones followed by n zeros is generated. The rotated word is then merged with the contents of the MQ register, under control of the generated mask. When bit 26 of register rB is a one, a mask of 32-n zeros followed by n ones is generated. A word of zeros is then merged with the contents of the MQ register, under control of the generated mask. The merged word is placed in register rA. The MQ register is not altered.

Shift Left with MQ
slq   rA,rS,rB

slq slq.
• **POWER** • **MPC601.**

Register rS is rotated left n bits where n is the shift amount specified in bits 27-31 of register rB. The rotated word is placed in the MQ register. When bit 26 of register rB is a zero, a mask of 32-n ones followed by n zeros is generated. When bit 26 of register rB is a one, a mask of all zeros is generated. The logical AND of the rotated word and the generated mask is placed into register rA.

Shift Left Word
slw   rA,rS,rB

slw slw.
• **32 bit PowerPC**

The contents of rS are shifted left the number of bits specified by rB[26-31]. Bits shifted out from position 0 are lost. Zeros are supplied to the empty positions on the right. The 32-bit result is placed into rA. If rB[26]=1, then rA is filled with zeros.

Shift Left Immediate
slwi   rA,rS,n

slwi slwi.
• **Alternative mnemomic**

This alternative will shift the contents of a register left by n bits while clearing any vacated bits. The value of n must be less than 32. It is equivalent of rlwinm rA,rS,n,0,31-n.

Shift Right Algebriac Double Word
srad   rA,rS,rB

srad srad.
• **64 bit PowerPC.**

Register rS is shifted right n bits where n is the shift amount specified by the seven low order bits of rB. All bits shifted from bit 63 are lost and bit 0 of rS is used to fill the vacated bit positions from the left. The result is then placed in rA. If any 1 bits are shifted or if rS is a negative number, then XER[CA] is set. If rB = 0 will cause rA to be set equal to rS and XER[CA] to be cleared.

If the shift value within rB is between 64 and 127, the result is zero(i.e. all the bits have been shifted and filled with zeros).

Shift Right Algebriac Double Word Immediate

sradi rA,rS,SH

sradi sradi.
• **64 bit PowerPC.**

Register rS is shifted right SH bits. All bits shifted from bit 63 are lost and bit 0 of rS is used to fill the vacated bit positions from the left. The result is then placed in rA. If any 1 bits are shifted or if rS is a negative number, then XER[CA] is set. If rB = 0 will cause rA to be set equal to rS and XER[CA] to be cleared.

Shift Right Algebraic Immediate with MQ

sraiq rA,rS,SH

sraiq sraiq.
• **POWER • MPC601.**

Register rS is rotated left 32-n bits where n is the shift amount specified by the operand SH. A mask of n zeros followed by 32-n ones is generated. The rotated word is placed in the MQ register. The rotated word is then merged with a word of 32 sign bits from register rS, under control of the generated mask. The merged word is placed in register rA. The rotated word is ANDed with the complement of the generated mask. This 32-bit result is ORed together and then ANDed with bit 0 of register rS to produce XER[CA] . Shift Right Algebraic instructions can be used for a fast divide by 2 to the power of n, if followed with addze.

Shift Right Algebraic with MQ

sraq rA,rS,rB

sraq sraq.
• **POWER • MPC601.**

Register rS is rotated left 32-n bits where n is the shift amount specified in bits 27-31 of register rB. When bit 26 of register rB is a zero, a mask of n zeros followed by 32-n ones is generated. When bit 26 of register rB is a one, a mask of all zeros is generated. The rotated word is placed in the MQ register. The rotated word is then merged with a word of 32 sign bits from register rS, under control of the generated mask. The merged word is placed in register rA. The rotated word is ANDed with the complement of the generated mask. This 32-bit result is ORed together and then ANDed with bit 0 of register rS to produce XER[CA]. This instruction can be used for a fast divide by 2 to the power of n if followed with addze.

Shift Right Algebraic Word
sraw   rA,rS,rB

sraw sraw.
• **32 bit PowerPC**

The contents of rS are shifted right the number of bits specified by rB[26-31]. The 32-bit result is placed into rA. XER[CA] is set to 1 if rS contains a negative number and any 1-bits are shifted out of position 31; otherwise XER[CA] is cleared to 0. An operand (rB) of zero causes rA to be loaded with the contents of rS, and XER[CA] to be cleared to 0. Condition register field CR0 is set based on the value written into rA.

If rB[26]=1, then rA is filled with 32 sign bits (bit 0) from rS.
If rB[26]=0, then rA is filled from the left with sign bits.

Shift Right Algebraic Word Immediate
srawi   rA,rS,SH

srawi srawi.
• **32 bit PowerPC**

The contents of rS are shifted right the number of bits specified by SH. Bits shifted out of position 31 are lost. The 32-bit result is then sign extended and placed into rA. XER[CA] is set if rS contains a negative number and any 1 bits are shifted out of position 31; otherwise XER(CA) is cleared. An operand SH of zero causes rA to be loaded with the contents of rS and XER[CA] to be cleared to 0.

Shift Right Double Word
srd   rA,rS,rB

srd srd.
• **64 bit PowerPC.**

Register rS is shifted right n bits where n i s the number of bits specified by the seven low order bits of rB. All bits shifted from bit 63 are lost and bit 0 of rS is used to fill the vacated bit positions from the left. The result is then placed in rA. If the shift value within rB is between 64 and 127, the result is zero(i.e. all the bits have been shifted and filled with zeros).

Shift Right Extended
sre   rA,rS,rB

sre sre.
• POWER • MPC601.

Register rS is rotated left 32-n bits where n is the shift amount specified in bits 27-31 of register rB. The rotated word is placed into the MQ register. A mask of n zeros followed by 32-n ones is generated. The logical AND of the rotated word and the generated mask is placed in register rA.

Shift Right Extended with MQ
sreq   rA,rS,rB

sreq sreq.
• POWER • MPC601.

Register rS is rotated left 32-n bits where n is the shift amount specified in bits 27-31 of register rB. A mask of n zeros followed by 32-n ones is generated. The rotated word is then merged with the contents of the MQ register, under control of the generated mask. The merged word is placed in register rA. The rotated word is placed into the MQ register.

Shift Right Immediate with MQ
sriq   rA,rS,SH

sriq sriq.
• POWER • MPC601.

Register rS is rotated left 32-n bits where n is the shift amount specified by operand SH. The rotated word is placed into the MQ register. A mask ot n zeros followed by 32-n ones is generated. The logical AND of the rotated word and the generated mask is placed in register rA.

Shift Right Long Immediate with MQ
srliq   rA,rS,SH

srliq srliq.
• POWER • MPC601.

Register rS is rotated left 32-n bits where n is the shift amount specified by operand SH. A mask of n zeros followed by 32-n ones is generated. The rotated word is then merged with the contents of the MQ register, under control of the generated mask. The merged word is placed in register rA. The rotated word is placed into the MQ register.

Shift Right Long with MQ
srlq   rA,rS,rB

srlq srlq.
• POWER • MPC601.

_____

Register rS is rotated left 32-n bits where n is the shift amount specified in bits 27-31 of register rB. When bit 26 of register rB is a zero, a mask of n zeros followed by 32-n ones is generated. The rotated word is then merged with the contents of the MQ register, under control of the generated mask. When bit 26 of register rB is a one, a mask of n ones followed by 32-n zeros is generated. A word of zeros is then merged with the contents of the MQ register, under control of the generated mask. The merged word is placed in register rA. The MQ register is not altered.

Shift Right with MQ
srq   rA,rS,rB

srq srq.
• POWER • MPC601.

_____

Register rS is rotated left 32-n bits where n is the shift amount specified in bits 27-31 of register rB. The rotated word is placed into the MQ register. When bit 26 of register rB is a zero, a mask of n zeros followed by 32-n ones is generated. When bit 26 of register rB is a one, a mask of all zeros is generated. The logical AND of the rotated word and the generated mask is placed in rA.

Shift Right Word
srw   rA,rS,rB

srw srw.
• 32 bit PowerPC

_____

The contents of rS are shifted right the number of bits specified by rB[26-31]. Bits shifted out from position 31 are lost. Zeros are supplied to the empty positions on the left. The 32-bit result is placed into rA.

If rB[26]=1, then rA is filled with zeros.

Shift Right Immediate
srwi   rA,rS,n

srwi srwi.
- **Alternative mnemomic**

This alternative will shift the contents of a register right by n bits while clearing any vacated bits. The value of n must be less than 32. It is equivalent of rlwinm rA,rS,32-n,0,31.

Store Byte
stb   rS,d(rA)

stb
- **32 bit PowerPC**

The effective address is the sum (rA|0)+d. Register rS[24-31] is stored into memory at the byte location addressed by the EA. The register rS is not altered as the result of this instruction.

Store Byte with Update
stbu   rS,d(rA)

stbu
- **32 bit PowerPC**

The effective address is the sum (rA|0)+d. rS[24 31] is stored into memory at the byte location addressed by the EA. The EA is placed into register rA to update it with the new address.

The MPC601 supports a version of this instruction where r0 is used as rA. This version is invalid within the PowerPC architecture.

Store Byte with Update Indexed
stbux   rS,rA,rB

stbux
- **32 bit PowerPC**

The effective address is the sum (rA|0)+(rB). rS[24-31] is stored into memory at the byte location addressed by the EA. The EA is placed into register rA to update it with the new address.

The MPC601 supports a version of this instruction where r0 is used as rA. This version is invalid within the PowerPC architecture.

Store Byte Indexed
stbx   rS,rA,rB

stbx
• **32 bit PowerPC**

The effective address is the sum (rA|0)+(rB). rS[24-31] is stored into the byte in memory addressed by the EA.

Store Double Word
std   rS,ds(rA)

std
• **64 bit PowerPC.**

The effective address is the sum (rA|0)+ (ds || 0b00). Register rS is stored into the double word in memory addressed by the EA.

Store Double Word
Conditional Indexed
stdcx.   rS,rA,rB

stdcx.
• **64 bit PowerPC.**

This instruction is used to implement a semaphore or other control structure and is used in conjunction with the ldarx instruction which creates the reservation in the first place.

The effective address is the sum (rA|0)+(rB). When the stdcx. is executed, its effective address is compared to the one stored with the reservation. If the reservation is still valid — another bus access has not been snooped by the processor — the stdcx. instruction will perform the store. If not the the store is not performed.

The EQ bit in the condition register field CR0 is modified to reflect whether the store operation was performed (i.e., whether a reservation existed when the stwcx. instruction began execution). If the store was completed successfully, the EQ bit is set to one. The EA must be a multiple of 8; otherwise, the alignment exception handler will be invoked if the word stored crosses a page boundary, or the results may be undefined.

Please note that this instruction must have the . suffix to ensure the the EQ bit is updated.

Store Double Word with Update

stdu   rS,ds(rA)

stdu
• **64 bit PowerPC.**

The effective address is the sum (rA)+ (ds‖0b00). Register rS is stored into the double word in memory addressed by the EA. The EA is then placed in rA to perform the update.

Store Double Word with Update Indexed

stdux   rS,rA,rB

stdux
• **64 bit PowerPC.**

The effective address is the sum (rA)+ (rB). Register rS is stored into the double word in memory addressed by the EA. The EA is then placed in rA to perform the update.

Store Double Word Indexed

stdx   rS,rA,rB

stdx
• **64 bit PowerPC.**

The effective address is the sum (rA)+ (rB). Register rS is stored into the double word in memory addressed by the EA.

Store Floating Point Double Precision

stfd   frS,d(rA)

stfd
• **32 bit PowerPC**

The effective address is the sum (rA|0)+d. The contents of register frS is stored into the double-word in memory addressed by the EA.

Store Floating Point Double Precision with Update

stfdu   frS,d(rA)

stfdu
• **32 bit PowerPC**

The effective address is the sum (rA|0)+d. The contents of register frS is stored into the double-word in memory addressed by the EA. The EA is placed into register rA to update it with the new address.

Store Floating Point Double Precision with Update Indexed

stfdux   frS,rA,rB

stfdux
• **32 bit PowerPC**

The EA is the sum (rA|0)+(rB). The contents of register frS is stored into the double-word in memory adressed by EA. The EA is placed into register rA.

The MPC601 supports a version of this instruction where r0 is used as rA. This version is invalid within the PowerPC architecture.

Store Floating Point Double Precision Indexed

stfdx   frS,rA,rB

stfdx
• **32 bit PowerPC**

The EA is the sum (rA|0)+(rB). The contents of register frS is stored into the double-word in memory addressed by the EA.

Store Floating Point as Integer Word Indexed

stfiwx   frS,rA,rB

stfiwx
• **Optional PowerPC(no MPC601 support).**

The EA is the sum (ra|0)+(rB) and specifies where in memory the low order 32 bits of the register frS is stored. The data is taken from bits 32 through to 63 of the register and are not converted in any way at all.

Store Floating Point Single Precision

stfs   frS,d(rA)

stfs

32 bit PowerPC

The effective address is the sum (rA|0)+d. The contents of register frS is converted to single-precision and stored into the double-word in memory addressed by the EA.

Store Floating Point Single Precision with Update

stfsu   frS,d(rA)

stfsu

32 bit PowerPC

The effective address is the sum (rA|0)+d. The contents of register frS is converted to single-precision and stored into the double-word in memory addressed by the EA. The EA is placed into register rA to update it with the new address.

Store Floating Point Single Precision with Update Indexed

stfsux   frS,rA,rB

stfsux

32 bit PowerPC

The EA is the sum (rA10)+d). The contents of register frS is converted to single-precision and stored into the word in memory addressed by the EA. The EA is placed into register rA to update with the new address.

The MPC601 supports a version of this instruction where r0 is used as rA. This version is invalid within the PowerPC architecture.

Store Floating Point Single Precision Indexed

stfsx   frS,rA,rB

stfsx

32 bit PowerPC

The EA is the sum (rA|0)+(rB). The contents of register frS is converted to single-precision and stored into the double-word in memory addressed by the EA.

Store Half word
sth   rS,d(rA)

sth

• 32 bit PowerPC

The effective address is the sum (rA|0)+d. rS[16-31] is stored into the half-word in memory addressed by the EA.

Store Half Word Byte-Reverse Indexed
sthbrx   rS,rA,rB

sthbrx

• 32 bit PowerPC

The effective address is the sum (rA|0)+(rB). rS[24-31] are stored into bits 0-7 of the half-word in memory addressed by the EA. rS[16-23] are stored into bits 8-15 of the half-word in memory addressed by the EA.

This instruction is used when accessing both little-endian and big-endian memory schemes. It effectively reverses the bytes that make up the half-word. If the half-word was in the register as AB it would be stored as BA. The individual bit settings within each byte are not changed.

Store Half word with Update
sthu   rS,d(rA)

sthu

• 32 bit PowerPC

The effective address is the sum (rA|0)+d. rS[16-31] is stored into the half-word in memory addressed by the EA. The EA placed into register rA to update it with the new address.

The MPC601 supports a version of this instruction where is used as rA. This is invalid within the PowerPC architecture.

Store Half word with Update Indexed
sthux   rS,rA,rB

sthux

• 32 bit PowerPC

The effective address is the sum (rA|0)+(rB). rS[16-31] stored into the half-word in memory addressed by the EA. The is placed into register rA to update it with the new address.The MPC601 supports a version of this instruction where r0 is used rA. This version is invalid within the PowerPC architecture.

Store Half word Indexed
sthx   rS,rA,rB

sthx
• 32 bit PowerPC

The effective address (EA) is the sum (rAI0)+(rB). rS[16-31] is stored into the half-word in memory addressed by the EA.

Store Multiple Word
stmw   rS,d(rA)

stmw
• 32 bit PowerPC

The effective address is the sum (rA|0)+d. If n=(32-rS). then n consecutive words starting at the EA are stored in the general purpose register file starting with rS and going through to r31. If rS=28 then 4 words are stored in memory. If rS=0, then the whole register file is stored out in memory.

If the EA is not a multiple of 4 the alignment exception handler may be invoked if a page boundary is crossed.

It is important to save the register contents prior to executing this instruction. Register corruption after the execution of this instruction is often due to the wrong count value using more registers than anticipated or programmed.

Store String Word Immediate
stswi   rS,rA,NB

stswi
• 32 bit PowerPC

The EA is (rA|0). NB is the number of bytes that will be stored in memory. If NB=0, the number is interpreted as 32 and not a byte count of zero. The bytes are stored from the registers, moving from left to right and starting with register rS and incrementing through the register file. The sequence will wrap around through r0 if required.

Store String Word Indexed
stswx   rS,rA,rB

stswx
• 32 bit PowerPC

The effective address is the sum (rA|0)+(rB). The number of bytes to be stored is contained in XER[25-31]. The bytes are stored from the registers, moving from left to right and starting with register rS and incrementing through the register file. The sequence will wrap around through r0 if required.

Store Word

stw   rS,d(rA)

stw

• 32 bit PowerPC

The effective address is the sum (rA|0)+d. Register rS is stored into the word in memory addressed by the EA.

Store Word Byte-Reverse Indexed

stwbrx   rS,rA,rB

stwbrx

• 32 bit PowerPC

The effective address is the sum (rA|0)+(rB). rS[24-31] are stored into bits 0-7 of the word in memory addressed by EA. Register rS[16-23] are stored into bits 8-15 of the word in memory addressed by the EA. Register rS[8-15] are stored into bits 16-23 of the word in memory addressed by the EA. rS[0-7] are stored into bits 24-31 of the word in memory addressed by the EA.

This instruction is used when accessing both little-endian and big-endian memory schemes. It effectively reverses the bytes that make up the word. If the word was in the register as ABCD, it would be stored as DCBA. The individual bit settings within each byte are not changed.

Store Word Conditional

stwcx.   rS,rA,rB

stwcx.

• 32 bit PowerPC

This instruction is used to implement a semaphore or other control structure and is used in conjunction with the lwar instruction which creates the reservation in the first place.

The effective address is the sum (rA|0)+(rB). When the stwcx. is executed, its effective address is compared to the one stored with the reservation. If the reservation is still valid – another bus access has not been snooped by the processor — the stwcx. instruction will perform the store. If not the the store is not performed.

The EQ bit in the condition register field CR0 is modified to reflect whether the store operation was performed (i.e., whether reservation existed when the stwcx. instruction began execution). If the store was completed successfully, the EQ bit is set to one. The EA must be a multiple of 4; otherwise, the alignment exception handler will be invoked if the word stored crosses page boundary, or the results may be undefined.

Please note that this instruction must have the . suffix to ensure the the EQ bit is updated.

Store Word with Update
stwu   rS,d(rA)

stwu

• **32 bit PowerPC**

_____

The effective address is the sum (rA|0)+d. Register rS is stored into the word in memory addressed by the EA. The EA is placed into register rA to update it with the new address.

The MPC601 supports a version of this instruction where r0 is used as rA. This version is invalid within the PowerPC architecture.

Store Word with Update Indexed
stwux   rS,rA,rB

stwux

• **32 bit PowerPC**

_____

The effective address is the sum (rA|0)+(rB). Register rS is stored into the word in memory addressed by the EA. The EA is placed into register rA to update it with the new address.

The MPC601 supports a version of this instruction where r0 is used as rA. This is invalid within the PowerPC architecture.

Store Word Indexed
stwx   rS,rA,rB

stwx

32 bit PowerPC

_____

The effective address is the sum (rA|0)+(rB). rS is stored into the word in memory addressed by the EA.

Subtract
sub   rD,rA,rB

sub

Alternative mnemomic

_____

The standard subf instruction substracts the second operand (rA) from the third (rB). This is an alternative mnemonic which has the more normal syntax where the third operand is subtracted from the secon. It is equivalent to subfc rD,rB,rA. It is equivalent to subf rD,rB,rA.

Subtract Carrying
subc rD,rA,rB

subc
• **Alternative mnemomic**

The standard subf instruction substracts the second operand (rA) from the third (rB). This is an alternative mnemomic which has the more normal syntax where the third operand is subtracted from the secon. It is equivalent to subfc rD,rB,rA.

Subtract from
subf rD,rA,rB

subf subf. subfo subfo.
• **32 bit PowerPC**

The sum –(rA) + (rB) +1 is placed into rD.

Subtract from Carrying
subfc rD,rA,rB

subfc subfc. subfco subfco.
• **32 bit PowerPC**

The sum –(rA) + (rB) + 1 is placed into register rD.

Subtract from Extended
subfe rD,rA,rB

subfe subfe. subfeo subfeo.
• **32 bit PowerPC**

The sum –(rA) + (rB) + XER(CA) is placed into register rD

Subtract from Immediate Carrying
subfic rD,rA,SIMM

subfic
• **32 bit PowerPC**

The sum (NOT(rA) + sign extended SIMM + 1) is placed in register rD.

Subtract from Minus One Extended
subfme rD,rA

subfme subfme. subfmeo subfmeo.
• 32 bit PowerPC

The sum NOT(rA) + XER(CA) + $FFFFFFFF is placed into register rD.

Subtract from Zero Extended
subfze rD,rA

subfze subfze. subfzeo subfzeo.
• 32 bit PowerPC

The sum NOT(rA) + XER(CA) is placed into register rD.

Subtract Immediate
subi rD,rA,value

subi
• Alternative mnemomic

This is an alternative meemonic and is equivalent to addi rD,rA,-value.

Subtract Immediate Carrying
subic rD,rA,value

subic
• Alternative mnemomic

This is an alternative mnemonic and is equivalent to addic rD,rA,-value.

Subtract Immediate Carrying with record
subic. rD,rA,value

subic.
• Alternative mnemomic

This is an alternative mnemonic and is equivalent to addic. rD,rA,-value.

Subtract Immediate Shifted
subis rD,rA,value

subis
• Alternative mnemomic

This is an alternative meemonic and is equivalent to addis rD,rA,-value.

Synchronize
sync

sync
• 32 bit PowerPC

This instruction is important when accessing I/O areas or other memory addresses where the strict program order must be enforced or completed before continuing the program execution. In short, it will ensure that all currently outstanding accesses to main memory before the execution of the eieio instruction are completed. Any subsequent memory accesses that are requested or initiated will be forced to wait. This synchronisation will order load and store operations to cache inhibited memory, and store operations to write through cache memory.

Previous accesses are deemed to have completed when they have completed to the point that they cannot cause an exception, all main memory accesses have been completed and their operation has been broadcast to comply with cache coherency policy.

This op code simply holds back the program execution to allow all the previous memory requests and operations to complete correctly. For example, if a program has issued several stores to an external peripheral to initialise and request it to perform a function, the eieio instruction can be used to ensure that all the stores have completed in order before reading the peripherals status register. Without, it is possible for the stores not to have completed before the read takes place because of the memory interface and its re-ordering capability, despite the program order.

Executing a sync instruction ensures that all instructions previously initiated by the processor appear to have completed before any subsequent instructions are initiated. When it completes, all memory accesses initiated by the given processor prior to the sync will have been performed with respect to all other mechanisms that access memory. The sync instruction can be used to ensure that the results of all stores into a data structure, performed in a 'critical section' of a program, are seen by other processors before the data structure is seen as unlocked. The Enforce In-Order Execution of I/0 (eieio) instruction may be more appropriate than sync for cases in which the only requirement is to control the order in which memory references are seen by I/O devices . With the MPC601, eieio performs in the same way as the sync instruction.

Trap Double Word
td TO,rA,rB

td

• **64 bit PowerPC.**

The contents of rA is compared with the contents of rB. The system trap handler is called if any of the following five conditions are met:

If sign-extended rA=A and sign-extended rB = B then
if (A < B & TO(0)) is true, take the trap exception.
if (A > B & TO(1)) is true, take the trap exception.
if (A = B & TO(2)) is true, take the trap exception.
If unsigned rA=A and unsigned rB = B then
if (A < B & TO(3)) is true, take the trap exception.
if (A > B & TO(4)) is true, take the trap exception.
As can be seen, the TO field effectively controls which comparisons are made.

Trap Double Word Immediate
tdi TO,rA,SIMM

tdi

• **64 bit PowerPC.**

TThe contents of rA is compared with the sign-extended SIMM operand. The system trap handler is called if any of the following five conditions are met:

If sign-extended rA=A and sign-extended SIMM = B then
if (A < B & TO(0)) is true, take the trap exception.
if (A > B & TO(1)) is true, take the trap exception.
if (A = B & TO(2)) is true, take the trap exception.
If unsigned rA=A and unsigned SIMM = B then
if (A < B & TO(3)) is true, take the trap exception.
if (A > B & TO(4)) is true, take the trap exception.
As can be seen, the TO field effectively controls which comparisons are made.

Translation Lookaside Buffer Invalidate All
tlbia

tlbia

• **Optional PowerPC(no MPC601 or MPC603 support).**

This instruction will invalidate the complete TLB, irrespective of the IT and DT bit settings within the MSR. This effectively clears all entries so that the TLB is empty.

This is a supervisor level instruction within the PowerPC architecture.

Translation Lookaside Buffer Invalidate All

tlbia

tlbia
• 64 bit PowerPC.

The complete translation lookaside buffer (TLB) is invalidated irrespective of the state of MSR[IR] and MSR[DR].

This is a supervisor level instruction and is optional within the PowerPC architecture.

Translation Lookaside Buffer Invalidate Entry

tlbie   rB

tlbie
• 32 bit PowerPC

The effective address is the contents of rB. The TLB search is done regardless of the settings of MSR[IT] and MSR[DT] but if the address is in an I/O controller segment, nothing further is done. If the TLB contains an entry for the EA, that entry is removed from the TLB, and a TLB invalidate operation is broadcast on the system bus. Care must be taken in multiprocessor systems to ensure coherency:

•   The tlbie instruction must be contained in a critical section of memory controlled by software locking, so that the tlbie is issued on only one processor at a time.

•   A sync instruction must be issued after every tlbie and at the end of the critical section. This causes hardware to wait for the effects of the preceding tlbie instructions) to propagate to all   processors. This ensures that all the TLBs on all the processors have been updated.

A processor detecting a TLB invalidate broadcast does the following:

1.   Prevents execution of any new load, store, cache control or tlbie instructions and prevents any new reference or change bit updates

2.   Waits for completion of any outstanding memory operations (including updates to the reference and change bits associated with the entry to be invalidated)

3.   Invalidates the two entries (both associativity classes) in the UTLB indexed by the matching address

4.   Resumes normal execution

Translation Lookaside Buffer Invalidate Entry by Index
tlbiex rB

tlbiex
• Optional PowerPC(no MPC601 or MPC603 support).

This instruction allows an individual TLB entry to be invalidated, irrepective of the IT and DT bit settings within the MSR. If rB contains the value N, then the Nth entry within the TLB will be invalidated. If this entry does not exist, the result is undefined and implementation dependent.

This is an optional, supervisor level instruction within the PowerPc architecture.

Please note that this instruction does not give any help or indication in determining which addresses are involved with a particular TLB entry. The use of this may require the development of special algorithms to track the translation process and thus deduce the TLB contents or alternatively, this instruction could be used in a random manner.

Translation Lookaside Buffer Invalidate Entry by Index
tlbiex rB

tlbiex
• Optional PowerPC(no MPC601 or MPC603 support).

This instruction allows an individual TLB entry to be invalidated, irrepective of the IT and DT bit settings within the MSR. If rB contains the value N, then the Nth entry within the TLB will be invalidated. If this entry does not exist, the result is undefined and implementation dependent.

This is an optional, supervisor level instruction within the PowerPc architecture.

Please note that this instruction does not give any help or indication in determining which addresses are involved with a particular TLB entry. The use of this may require the development of special algorithms to track the translation process and thus deduce the TLB contents or alternatively, this instruction could be used in a random manner.

Translation Lookaside Buffer Synchronise

tlbsync

tlbsync

• **Optional PowerPC(no MPC601 or MPC603 support).**

This instruction synchronises the translation lookaside buffers by forcing the processor to wait until all previous tlbie, tlbiex and tlbia instructions have completed. The PowerPC architecture defines this instruction as essential if either eciwx or ecowx instructions are supported or if a TLB invalidation instruction causes an external broadcast. The MPC601 does do this but does not support this instruction.

This is a supervisor only instruction.

Trap Word

tw TO,rA,rB

tw

• **32 bit PowerPC**

The contents of rA is compared with the contents of rB. The system trap handler is called if any of the following five conditions are met:

If sign-extended rA=A and sign-extended rB = B then

if (A < B & TO(0)) is true, take the trap exception.
if (A > B & TO(1)) is true, take the trap exception.
if (A = B & TO(2)) is true, take the trap exception.

If unsigned rA=A and unsigned rB = B then

if (A < B & TO(3)) is true, take the trap exception.
if (A > B & TO(4)) is true, take the trap exception.

As can be seen, the TO field effectively controls which comparisons are made.

Trap if Equal

tweq rA,rB

tweq

• **Alternative mnemomic**

This is an alternative mnemomic and use the Trap Word instruction with no TO field. The condition is defined by the mnemomic itself.

Trap if Equal
tweqi   rA,SIMM

tweqi
• **Alternative mnemomic**

This is an alternative mnemomic and use the Trap Word Immediate instruction with no TO field. The condition is defined by the mnemomic itself.

Trap if Greater Than or Equal
twge   rA,rB

twge
• **Alternative mnemomic**

This is an alternative mnemomic and use the Trap Word instruction with no TO field. The condition is defined by the mnemomic itself.

Trap if Greater Than or Equal
twgei   rA,SIMM

twgei
• **Alternative mnemomic**

This is an alternative mnemomic and use the Trap Word Immediate instruction with no TO field. The condition is defined by the mnemomic itself.

Trap if Greater Than
twgt   rA,rB

twgt
• **Alternative mnemomic**

This is an alternative mnemomic and use the Trap Word instruction with no TO field. The condition is defined by the mnemomic itself.

Trap if Greater Than
twgti   rA,SIMM

twgti
• **Alternative mnemomic**

This is an alternative mnemomic and use the Trap Word Immediate instruction with no TO field. The condition is defined by the mnemomic itself.

Trap Word Immediate
twi TO,rA,SIMM

twi

• **32 bit PowerPC**

The contents of rA is compared with the sign-extended SIMM operand. The system trap handler is called if any of the following five conditions are met:

If sign-extended rA=A and sign-extended SIMM = B then
if (A < B & TO(0)) is true, take the trap exception.
if (A > B & TO(1)) is true, take the trap exception.
if (A = B & TO(2)) is true, take the trap exception.
If unsigned rA=A and unsigned SIMM = B then
if (A < B & TO(3)) is true, take the trap exception.
if (A > B & TO(4)) is true, take the trap exception.

As can be seen, the TO field effectively controls which comparisons are made.

Trap if Less Than or Equal
twle rA,rB

twle

• **Alternative mnemomic**

This is an alternative mnemomic and use the Trap Word instruction with no TO field. The condition is defined by the mnemomic itself.

Trap if Less Than or Equal
twlei rA,SIMM

twlei

• **Alternative mnemomic**

This is an alternative mnemomic and use the Trap Word Immediate instruction with no TO field. The condition is defined by the mnemomic itself.

Trap if Logically Greater Than or Equal
twlge rA,rB

twlge

• **Alternative mnemomic**

This is an alternative mnemomic and use the Trap Word instruction with no TO field. The condition is defined by the mnemomic itself.

Trap if Logically Greater Than or Equal
twlgei   rA,SIMM

twlgei
• **Alternative mnemomic**

This is an alternative mnemomic and use the Trap Word Immediate instruction with no TO field. The condition is defined by the mnemomic itself.

Trap if Logically Greater Than
twlgt   rA,rB

twlgt
• **Alternative mnemomic**

This is an alternative mnemomic and use the Trap Word instruction with no TO field. The condition is defined by the mnemomic itself.

Trap if Logically Greater Than
twlgti   rA,SIMM

twlgti
• **Alternative mnemomic**

This is an alternative mnemomic and use the Trap Word Immediate instruction with no TO field. The condition is defined by the mnemomic itself.

Trap if Logically Less Than or Equal
twlle   rA,rB

twlle
• **Alternative mnemomic**

This is an alternative mnemomic and use the Trap Word instruction with no TO field. The condition is defined by the mnemomic itself.

Trap if Logically Less Than or Equal
twllei   rA,SIMM

twllei
• **Alternative mnemomic**

This is yet another alternative mnemomic and uses the Trap Word Immediate instruction with no TO field. The condition to be tested is defined by the mnemomic itself(logically less than or equal).

Trap if Logically Less Than
twllt   rA,rB

twllt
• **Alternative mnemomic**

This is an alternative mnemomic and use the Trap Word instruction with no TO field. The condition is defined by the mnemomic itself.

Trap if Logically Less Than
twllti   rA,SIMM

twllti
• **Alternative mnemomic**

This is an alternative mnemomic and use the Trap Word Immediate instruction with no TO field. The condition is defined by the mnemomic itself.

Trap if Logically Not Greater Than
twlng   rA,rB

twlng
• **Alternative mnemomic**

This is an alternative mnemomic and use the Trap Word instruction with no TO field. The condition is defined by the mnemomic itself.

Trap if Logically Not Greater Than
twlngi   rA,SIMM

twlngi
• **Alternative mnemomic**

This is an alternative mnemomic and use the Trap Word Immediate instruction with no TO field. The condition is defined by the mnemomic itself.

Trap if Logically Not Less Than
twlnl   rA,rB

twlnl
• **Alternative mnemomic**

This is an alternative mnemomic and use the Trap Word instruction with no TO field. The condition is defined by the mnemomic itself.

Trap if Logically Not Less Than
twlnli   rA,SIMM

twlnli
• **Alternative mnemomic**

This is an alternative mnemomic and use the Trap Word Immediate instruction with no TO field. The condition is defined by the mnemomic itself.

Trap if Less Than
twlt   rA,rB

twlt
• **Alternative mnemomic**

This is an alternative mnemomic and use the Trap Word instruction with no TO field. The condition is defined by the mnemomic itself.

Trap if Less Than
twlti   rA,SIMM

twlti
• **Alternative mnemomic**

This is an alternative mnemomic and use the Trap Word Immediate instruction with no TO field. The condition is defined by the mnemomic itself.

Trap if Not Equal
twne   rA,rB

twne
• **Alternative mnemomic**

This is an alternative mnemomic and use the Trap Word instruction with no TO field. The condition is defined by the mnemomic itself.

Trap if Not Equal
twnei   rA,SIMM

twnei
• **Alternative mnemomic**

This is an alternative mnemomic and use the Trap Word Immediate instruction with no TO field. The condition is defined by the mnemomic itself.

Trap if Not Greater Than
twng   rA,rB

twng
• **Alternative mnemomic**

This is an alternative mnemomic and use the Trap Word instruction with no TO field. The condition is defined by the mnemomic itself.

Trap if Not Greater Than
twngi   rA,SIMM

twngi
• **Alternative mnemomic**

This is an alternative mnemomic and use the Trap Word Immediate instruction with no TO field. The condition is defined by the mnemomic itself.

Trap if Not Less Than
twnl   rA,rB

twnl
• **Alternative mnemomic**

This is an alternative mnemomic and use the Trap Word instruction with no TO field. The condition is defined by the mnemomic itself.

Trap if Not Less Than
twnli   rA,SIMM

twnli
• **Alternative mnemomic**

This is an alternative mnemomic and use the Trap Word Immediate instruction with no TO field. The condition is defined by the mnemomic itself.

XOR
xor   rA,rS,rB

xor xor.
• **32 bit PowerPC**

The contents of rS is XORed with the contents of rB and the result is placed into register rA.

XOR Immediate
xori   rA,rS,UIMM

xori
• **32 bit PowerPC**

The contents of rS is XORed with $0000 || UIMM and the result is placed into rA.

XOR Immediate Shifted
xoris   rA,rS,UIMM

xoris
• **32 bit PowerPC**

The contents of rS is XORed with UIMM || $0000 and the result is placed into rA.

Appendix A

Power architecture instructions

The following instructions are 32 bit Power architecture instructions and are only supported on the MPC601 and not on the other PowerPC processors. For complete compatibility and portability, these instructions should not be used.

Op code	Description
abs	Absolute
clcs	Cache Line Compute Size
doz	Difference or Zero
dozi	Difference or Zero Immediate
div	Divide
divs	Divide Short
lscbx	Load String and Compare Byte Indexed
maskg	Mask Generate
maskir	Mask Insert from Register
mul	Multiply
nabs	Negative absolute
rlmi	Rotate Left then Mask Insert
rrib	Rotate Right and Insert Bit
sle	Shift Left Extended
sleq	Shift Left Extended with MQ
sliq	Shift Left immediate with MQ
slliq	Shift Left Long Immediate with MQ
sllq	Shift Left Long with MQ
slq	Shift Left with MQ
sraiq	Shift Right Algebraic Immediate with MQ
sraq	Shift Right Algebraic with MQ
sre	Shift Right Extended
sreq	Shift Right Extended with MQ
sriq	Shift Right Immediate with MQ
srliq	Shift Right Long Immediate with MQ
srlq	Shift Right Long with MQ
srq	Shift Right with MQ

Appendix B

64 bit PowerPC instructions

The following instructions are 64 bit PowerPC instructions and are not supported on the 32 bit implementations such as the MPC601, MPC603 and MPC604. The MPC620 can support them if it is setup in 64 bit mode.

Op code	Description
cntlzd	Count leading zero double word
extsw	Extend Sign Word
fcfid	Floating Point Convert from Integer Double Word
fctidz	Floating Point Convert to Integer DoubleWord with Round toward Zero
fsqrt	Floating Square Root
fsqrts	Floating Square Root Single
ld	Load Double Word
ldarx	Load Double Word and Reserve Indexed
ldu	Load Double Word with Update
ldux	Load Double Word with Update Indexed
ldx	Load Double Word Indexed
lwa	Load Word Algebraic
lwaux	Load Word Algebriac with Update Indexed
lwax	Load Word Algebriac Indexed
mulhd	Multiply High Double Word
mulhdu	Multiply High Double Word Unsigned
mulld	Multiply Low Double Word
rldcl	Rotate Left Double Word then Clear Left
rldcr	Rotate Left Double Word then Clear Right
rldic	Rotate Left Double Word Immediate then Clear
rldicl	Rotate Left Double Word Immediate then Clear Left
rldicr	Rotate Left Double Word Immediate then Clear Right
rldimi	Rotate Left Double Word Immediate then Mask Insert
slbia	SLB Invalidate All
slbie	SLB Invalidate Entry
sld	Shift Left Double Word
srad	Shift Right Algebraic Double Word
sradi	Shift Right Algebriac Double Word Immediate
srd	Shift Right Double Word
std	Store Double Word
stdcx.	Store Double Word Conditional Indexed
stdu	Store Double Word with Update
stdux	Store Double Word with Update Indexed
stdx	Store Double Word Indexed
td	Trap Double Word

Op code	Description
tdi	Trap Double Word Immediate
tlbia	Translation Lookaside Buffer Invalidate All
divd	Divide Double Word
divdu	Divide Double Word Unsigned

Appendix C

Alternative mnemomics

The following mnemomics are classed as alternatives and as such do not exist as actual instructions.

Op code	Description
subi	Subtract Immediate
subis	Subtract Immediate Shifted
subic	Subtract Immediate Carrying
subic.	Subtract Immediate Carrying with record
subc	Subtract Carrying
sub	Subtract
cmpwi	Compare Word Immediate
cmpw	Compare Word
cmplw	Compare Logical Word
cmplwi	Compare Word Logical Immediate
extlwi	Extract and Left Justify Immediate
extrwi	Extract and Right Justify Immediate
inslwi	Insert from Left Immediate
insrwi	Insert from Right Immediate
rotlwi	Rotate Left Immediate
rotrwi	Rotate Right Immediate
rotrwi	Rotate Left
slwi	Shift Left Immediate
srwi	Shift Right Immediate
clrrwi	Clear Right Immediate
clrlwi	Clear Left Immediate
clrlslwi	Clear Left and Shift Left Immediate
crset	Condition Register Set
crclr	Condition Register Clear
crmove	Condition Register Move
crnot	Condition Register Not
twlti	Trap if Less Than
twle	Trap if Less Than or Equal
tweqi	Trap if Equal
twgei	Trap if Greater Than or Equal
twgti	Trap if Greater Than
twnli	Trap if Not Less Than
twnei	Trap if Not Equal
twngi	Trap if Not Greater Than
twllti	Trap if Logically Less Than
twllei	Trap if Logically Less Than or Equal
twlgei	Trap if Logically Greater Than or Equal
twlgti	Trap if Logically Greater Than
twlnli	Trap if Logically Not Less Than
twlngi	Trap if Logically Not Greater Than
twlt	Trap if Less Than
tweq	Trap if Equal
twge	Trap if Greater Than or Equal
twgt	Trap if Greater Than
twlge	Trap if Logically Greater Than or Equal

Op code	Description
twlgt	Trap if Logically Greater Than
twlle	Trap if Logically Less Than or Equal
twllt	Trap if Logically Less Than
twlng	Trap if Logically Not Greater Than
twlnl	Trap if Logically Not Less Than
twne	Trap if Not Equal
twng	Trap if Not Greater Than
twnl	Trap if Not Less Than
twlei	Trap if Less Than or Equal
mtxer	Move to XER
mfxer	Move from XER
mtlr	Move to Link Register
mtctr	Move to Count Register
mtdsisr	Move to DSISR
mtdar	Move to Data Address Register
mtdec	Move to Decrementer Register
mtsdr1	Move to SDR1
mtsrr0	Move to Status Save Restore Register 0
mtsrr1	Move to Status Save Restore Register 1
mtsprg	Move to Special Purpose Register n
mtasr	Move to Address Space Register
mtear	Move to External Address Register
mttbl	Move to Time Base Lower register
mttbu	Move to Time Base Upper register
mtibatu	Move to IBAT Upper register
mtibatl	Move to IBAT Lower register
mtdbatu	Move to DBAT Upper register
mtdbatl	Move to DBAT Lower register
mflr	Move from Link Register
mfctr	Move from Count Register
mfdsisr	Move from DSISR Register
mfdar	Move from Data Address Register
mfdec	Move from Decrementer Register
mfsdr1	Move from SDR1 Register
mfsrr0	Move from Status Save Restore 0 Register
mfsrr1	Move from Status Save Restore 1 Register
mfsprg	Move from Special Purpose Register n
mfasr	Move from Address Space Register
mfear	Move from External Access Register
mftbl	Move from Time Base Lower Register
mftbu	Move from Time Base Upper Register
mfpvr	Move from Processor Version Register
mfibatu	Move from IBAT Upper Register
mfdbatu	Move from DBAT Upper Register
mfibatl	Move from IBAT Lower Register
mfdbatl	Move from DBAT Lower Register
nop	No-op
li	Load Immediate
lis	Load ImmediateSigned
la	Load Address
mr	Move Register
not	Complement Register

Appendix D
M68000 to PowerPC
instruction reference

This appendix provides a quick and easy cross-reference for programmers that are either moving from the M68000 environment or are in the process of converting M68000 assembler code to PowerPC. The MMU instructions have not been included because they are hardware specific and thus are not supported in the PowerPC architecture. The floating point instructions have not been added because the PowerPC support is simply limited to essentially the basic +,–,x and / operations and does not support the transcendental functions that the MC68881 and MC68882 floating point units provide.

M68000 instruction	PowerPC equivalent
ABCD Add BCD	No direct support
ADD Add	Use a suitable add instruction
ADDA Add address	Use a suitable add instruction
ADDI Add immediate	Use add immediate if immediate value is not greater than 16 bits. Use a register to create a temporary immediate value if a larger immediate value is needed.
ADDQ Add quick	Use add immediate.
ADDX Add with extend	Use a suitable add instruction.
AND Logical AND	Use the and instruction.
ANDI Logical AND immediate	Use and immediate instruction.
ASL Arithmetic shift left	Use an appropriate rotate or shift left instruction to simulate the operation.
ASR Arithmetic shift right	Use an appropriate rotate or shift left instruction to simulate the operation.
Bcc Branch if cc is true	Use an appropriate branch instruction taking into account specultaive execution.
BCHG Test bit and change	No direct support. Requires simulation using the reservation station technique to provide the atomic operation.
BCLR Test bit and clear	No direct support. Requires simulation using the reservation station technique to provide the atomic operation.
BFCHG Test bit field and change	No direct support. Requires simulation using the reservation station technique to provide the atomic operation.

M68000 instruction	**PowerPC equivalent**
BFCLR Test bit field and clear	
	No direct support. Requires simulation using the reservation station technique to provide the atomic operation.
BFEXTS Extract signed bit field	
	Use an appropriate rotate or shift instruction to simulate the operation.
BFEXTU Extract unsigned bit field	
	Use an appropriate rotate or shift left instruction to simulate the operation.
BFFFO Bit field find first 1	Use an appropriate rotate or shift left instruction(s) to simulate the operation. Count leading zeros can also be used as part of a simulation.
BFINS Insert bit field	Use an appropriate rotate or shift left instruction(s) to simulate the operation.
BFSET Test bit field and set	No direct support. Requires simulation using the reservation station technique to provide the atomic operation.
BFTST Test bit field	No direct support. Requires simulation using the reservation station technique to provide the atomic operation.
BKPT Breakpoint	No direct support. Use the breakpoint registers in the processor supervisor model to set breakpoints.
BRA Branch	Use Branch alwasy instruction.
BSET Test bit and set	No direct support. Requires simulation using the reservation station technique to provide the atomic operation.
BSR Branch to subroutine	Use a suitable branch instruction that uses the link register. The choice will depend on how far the target address. The link register contents prior to the branch may need saving to preserve the context of any previous brnaches. The system call instruction can also be used in some circumstances.
BTST Bit test	No direct support. Requires simulation using the reservation station technique to provide the atomic operation.
CALLM Call module	No direct support

M68000 instruction	PowerPC equivalent

CAS Compare and swap operands

No direct support. Requires simulation using the reservation station technique to provide the atomic operation.

CAS2 Compare and swap dual operands

No direct support. Requires simulation using the reservation station technique to provide the atomic operation.

CHK Check register against bounds

No direct support. Requires simulation using the reservation station technique to provide the atomic operation.

CHK2 Check with upper/lower bounds

No direct support. Requires simulation using the reservation station technique to provide the atomic operation.

CLR Clear

Use r0 within an add or logical operation to write $00000000 into the target register. If the location is in external memory, use a store instruction with r0 instead.

CMP Compare

No direct support. Requires simulation using the reservation station technique to provide the atomic operation.

CMP2 Compare with upper/lower bounds

No direct support. Requires simulation using the reservation station technique to provide the atomic operation.

CMPA Compare address   Use a suitable compare instruction.

CMPI Compare immediate   Use a suitable compare instruction.

CMPM Compare memory with memory

No direct support. Requires simulation using the reservation station technique to provide the atomic operation.

DBcc Test,decrement, branch if cc true   Use branch instructions that decrement the CTR register to simulate.

DIVS Signed divide

Use divide instructions but remember the PowerPC instructions do not create remainders directly.

DIVSL Signed divide

Use divide instructions but remember the PowerPC instructions do not create remainders directly.

M68000 instruction	PowerPC equivalent
DIVU Unsigned divide	Use divide instructions but remember the PowerPC instructions do not create remainders directly.
DIVUL Unsigned divide	Use divide instructions but remember the PowerPC instructions do not create remainders directly.
EOR Logical EXCLUSIVE OR	
	Use the XOR instruction.
EORI Logical EXCLUSIVE OR	
	Use the XORI instruction.
EXG Exchange registers	No direct support. Use a rotate instruction.
EXT Sign extend	Use an extend instruction.
EXTB Sign extend	Use the extend byte instruction.
ILLEGAL Illegal opcode trap	
	Illegal op codes will be trapped out by the PowerPC processors. As such they is no defined illegal instruction — any invalid instruction will do.
JMP Jump	Use a branch always instruction.
LEA Load effective address	Use an add immediate instruction.
LINK Link and allocate	No direct support. Manipulate address pointers directly.
MOVE CCR Move condition code	
	Use a CCR instruction to move the contents of a condition field to another. Use a move to or move from condition register instruction.
MOVEA Move address	Use an add or logical instruction to move the register contents to another register. A store instruction will be needed to move the contents to memory.
MOVE SR Move to/from status register	
	Use the move to and from MSR instructions.
MOVE USP Move user stack pointer	
	Use an add or logical instruction to move the register contents to another register. A store instruction will be needed to move the contents to memory.
MOVEC Move control register	
	Use an appropriate move to and move from instruction e.g. move to special purpose register.
MOVEM Move multiple registers	
	Use store multiple words to move registers to memory. Use the Load multiple words to restore them.

M68000 instruction	PowerPC equivalent
MOVEP Move peripheral data	
	No direct support. Simulate using the reservation station technique with mit manipulation.
MOVEQ Move quick	Use an add or logical immediate instruction to simulate.
MOVES Move alternate address space	
	No direct support.
MULS Signed multiply	Use the multiply high and low instructions to simulate the multiply operation.
MULU Unsigned multiply	Use the multiply high and low instructions to simulate the multiply operation.
NBCD Negate BCD with extend	
	No direct support.
NEG Negate	No direct support.
NEGX Negate with extend	No direct support.
NOP No operation	A logical operation that use r0 to do nothing will simulate a no-op but if sunchronisation of the processor is required use the isync or sync instructions.
NOT Logical complement	Use nand instruction.
OR Logical OR	Use the or instruction.
ORI Logical OR immediate	Use the or immediate instruction.
PACK Pack BCD data	No direct support
PEA Push effective address	No direct support. Manipulate addresses directly.
RESET Reset external devices	
	No direct support.
ROL Rotate left	Use an appropriate rotate instruction.
ROR Rotate right	Use an appropriate rotate instruction.
ROXL Rotate left with extend	
	Use an appropriate rotate instruction.
ROXR Rotate right with extend	
	Use an appropriate rotate instruction.
RTE Return from exception	Use the rfi instruction.
RTM Return from module	No direct support.
RTR Return and restore codes	
	No direct support.
RTS Return from subroutine	Use the branch always with the destination address in the link register.
SBCD Subtract BCD	No direct support.
Scc Set conditionally	No direct support.
STOP Stop	No direct support.

SUB Subtract	Use an appropriate subtract instruction.
SUBA Subtract address	Use an appropriate subtract instruction.
SUBI Subtract immediate	Use an appropriate subtract instruction.
SUBQ Subtract quick	Use an appropriate subtract instruction.
SUBX Subtract with extend	Use an appropriate subtract instruction.
SWAP Swap register words	No direct support. Use a rotate instruction to simulate.
TAS Test operand and set	No direct support. Requires simulation using the reservation station technique to provide the atomic operation.
TRAP Trap	No direct support.
TRAPcc Trap conditionally	No direct support.
TRAPV Trap on overflow	No direct support.
TST Test operand	No direct support. Requires simulation using the reservation station technique to provide the atomic operation.
UNLK Unlink	No direct support. Manipulate addresses directly.
UNPK Unpack BCD data	No direct support.

Appendix E
Special purpose registers

This appendix gives the SPR numbering for the special purpose registers that are supported within the MPC601. MPC603 and MPC604 programming models.

MPC601 programming model

User-Level SPRs

Number	Register
SPR0	MQ Register
SPR1	XER — Integer Exception Register
SPR4	RTCU — RTC Upper Register 1 3
SPR5	RTCL — RTC Lower Register 1'3
SPR8	LR — Link Register
SPR9	CTR — Count Register

Supervisor-Level SPRs

Number	Register
SPR18	DSISR — DAE/ Source Instruction Service Register
SPR19	DAR — Data Address Register
SPR22	DEC — Decrementer Register
SPR25	SDR1 — Table Search Descriptor Register 1
SPR26	SRR0 — Save and Restore Register 0
SPR27	SRR1 — Save and Restore Register 1
SPR272	SPRG0 — SPR General 0
SPR273	SPRG1 — SPR General 1
SPR274	SPRG2 — SPR General 2
SPR275	SPRG3 — SPR General 3
SPR282	EAR — External Access Register
SPR286	PVR — Processor Version Register
SPR528	BAT0U—Instruction BAT 0 Upper
SPR529	BAT0L—Instruction BAT 0 Lower
SPR530	BAT1U—Instruction BAT 1 Upper
SPR531	BAT1 L—Instruction BAT 1 Lower
SPR532	BAT2U—Instruction BAT 2 Upper
SPR533	BAT2L—Instruction BAT 2 Lower
SPR534	BAT3U—Instrucbon BAT 3 Upper
SPR535	BAT3L—Instruction BAT 3 Lower
SPR1008	HID0
SPR1009	HID1
SPR1010	HID2 (IABR)
SPR1013	HID5 (DABR)
SPR1023	HID15 (PIR)

MPC603 programming model

User-Level SPRS

Number	Register
SPR1	XER — Integer Exception Register
SPR8	LR—Link Register
SPR9	CTR—Count Register

Supervisor-Level SPRS

Number	Register
SPR18	DSISR—DAE/ Source Instruction Service Register
SPR19	DAR—Data Address Register
SPR22	DEC—Decrementer Register
SPR25	SDR1—Table Search Description Register 1
SPR26	SRR0—Save and Restore Register 0
SPR27	SRR1—Save and Restore Register 1
SPR272	SPRG0—SPR General 0
SPR273	SPRG1—SPR General 1
SPR274	SPRG2—SPR General 2
SPR275	SPRG3—SPR General 3
SPR282	EAR—External Access Register
SPR284	TBL—Time Base
SPR285	TBU—Time Base Upper
SPR287	PVR—Processor Version Register
SPR528	IBAT0U—IBAT 0 Upper
SPR529	IBAT0L—IBAT 0 Lower
SPR530	IBAT1U—IBAT 1 Upper
SPR531	IBAT1 L—IBAT 1 Lower
SPR532	IBAT2U—IBAT 2 Upper
SPR533	IBAT2L—IBAT 2 Lower
SPR534	IBAT3U—IBAT 3 Upper
SPR535	IBAT3L—IBAT 3 Lower
SPR536	DBAT0U—DBAT 0 Upper
SPR537	DBAT0L—DBAT 0 Lower
SPR538	DBAT1 U—DBAT 1 Upper
SPR539	DBAT1 L—DBAT 1 Lower
SPR540	DBAT2U—DBAT 2 Upper
SPR541	DBAT2L—DBAT 2 Lower
SPR542	DBAT3U—DBAT 3 Upper
SPR543	DBAT3L—DBAT 3 Lower
SPR976	DM ISS—Data TLB Miss Address Register
SPR977	DCMP—Data TLB Miss Compare Register
SPR978	HASH1—PTEG1 Address Register
SPR979	HASH2—PTEG2 Address Register
SPR980	IMISS—Instruction TLB Miss Address Register
SPR981	ICMP—Instruction TLB Miss Compare Register
SPR982	RPA—Real Page Address Register
SPR1008	HID0—Hardware Implementation Register
SPR1010	HID2—Instruction Address Breakpoint (IABR)

MPC604 programming model

User-Level SPRS

Number	Register
SPR1	XER — Integer Exception Register
SPR8	LR—Link Register
SPR9	CTR—Count Register

Supervisor-Level SPRS

Number	Register
SPR18	DSISR—DAE/ Source Instruction Service Register
SPR19	DAR—Data Address Register
SPR22	DEC—Decrementer Register
SPR25	SDR1—Table Search Description Register 1
SPR26	SRR0—Save and Restore Register 0

Number	Register
SPR27	SRR1—Save and Restore Register 1
SPR272	SPRG0—SPR General 0
SPR273	SPRG1—SPR General 1
SPR274	SPRG2—SPR General 2
SPR275	SPRG3—SPR General 3
SPR282	EAR—External Access Register
SPR284	TBL—Time Base
SPR285	TBU—Time Base Upper
SPR287	PVR—Processor Version Register
SPR528	IBAT0U—IBAT 0 Upper
SPR529	IBAT0L—IBAT 0 Lower
SPR530	IBAT1U—IBAT 1 Upper
SPR531	IBAT1 L—IBAT 1 Lower
SPR532	IBAT2U—IBAT 2 Upper
SPR533	IBAT2L—IBAT 2 Lower
SPR534	IBAT3U—IBAT 3 Upper
SPR535	IBAT3L—IBAT 3 Lower
SPR536	DBAT0U—DBAT 0 Upper
SPR537	DBAT0L—DBAT 0 Lower
SPR538	DBAT1 U—DBAT 1 Upper
SPR539	DBAT1 L—DBAT 1 Lower
SPR540	DBAT2U—DBAT 2 Upper
SPR541	DBAT2L—DBAT 2 Lower
SPR542	DBAT3U—DBAT 3 Upper
SPR543	DBAT3L—DBAT 3 Lower
SPR952	MMCR0—Monitor Mode Control Register 0
SPR953	PMC1—Performance Monitor Counter Register 1
SPR954	PMC2—Performance Monitor Counter Register 2
SPR955	SIA—Sampled Instruction Address
SPR959	SDA—Sampled Data Address
SPR1010	IABR—Instruction Address Breakpoint Register
SPR1013	DABR—Data Address Breakpoint Register
SPR1008	HID0—Hardware Implementation Register
SPR1023	PIR—Processor Identification Register

Index

I

L

M

S